T0234084

Fixed-Point Signal Processing

Synthesis Lectures on Signal Processing

Editor
José Moura, *Carnegie Mellon University*

© Springer Nature Switzerland AG 2022
Reprint of original edition © Morgan & Claypool 2009

All rights reserved. No part of this publication may be reproduced, stored in a retrieval system, or transmitted in any form or by any means—electronic, mechanical, photocopy, recording, or any other except for brief quotations in printed reviews, without the prior permission of the publisher.

Fixed-Point Signal Processing
Wayne T. Padgett and David V. Anderson

ISBN: 978-3-031-01405-5   paperback
ISBN: 978-3-031-02533-4   ebook

DOI 10.1007/978-3-031-02533-4

A Publication in the Springer series
SYNTHESIS LECTURES ON SIGNAL PROCESSING

Lecture #9
Series Editor: José Moura, *Carnegie Mellon University*

Series ISSN
Synthesis Lectures on Signal Processing
Print 1932-1236   Electronic 1932-1694

Fixed-Point Signal Processing

Wayne T. Padgett
Rose-Hulman Institute of Technology

David V. Anderson
Georgia Institute of Technology

SYNTHESIS LECTURES ON SIGNAL PROCESSING #9

ABSTRACT

This book is intended to fill the gap between the "ideal precision" digital signal processing (DSP) that is widely taught, and the limited precision implementation skills that are commonly required in fixed–point processors and field programmable gate arrays (FPGAs). These skills are often neglected at the university level, particularly for undergraduates. We have attempted to create a resource both for a DSP elective course, and for the practicing engineer with a need to understand fixed–point implementation. Although we assume a background in DSP, Chapter 2 contains a review of basic theory, and Chapter 3 reviews random processes to support the noise model of quantization error. Chapter 4 details the binary arithmetic that underlies fixed–point processors, and then introduces fractional format for binary numbers. Chapter 5 covers the noise model for quantization error and the effects of coefficient quantization in filters. Because of the numerical sensitivity of IIR filters, they are used extensively as an example system in both Chapters 5 and 6. Fortunately, the principles of dealing with limited precision can be applied to a wide variety of numerically sensitive systems, not just IIR filters. Chapter 6 discusses the problems of product roundoff error, and various methods of scaling to avoid overflow. Chapter 7 discusses limit cycle effects and a few common methods for minimizing them.

There are a number of simple exercises integrated into the text to allow you to test your understanding. Answers to the exercises are included in the footnotes. A number of Matlab examples are provided in the text. They generally assume access to the Fixed–Point Toolbox. If you lack access to this software, consider either purchasing or requesting an evaluation license from The Mathworks. The code listed in the text and other helpful Matlab code is also available at `http://www.morganclaypool.com/page/padgett` and `http://www.rose-hulman.edu/~padgett/fpsp`. You will also find Matlab exercises designed to demonstrate each of the four types of error discussed in Chapters 5 and 6. Simulink examples are also provided on the web site.

KEYWORDS

binary arithmetic, clipping, coefficient quantization, digital signal processing (DSP), field programmable gate array (FPGA), fixed–point, IIR filter, limit cycles, Matlab, numerical sensitivity, overflow, $Q15$, quantization error, quantization noise, roundoff noise, saturation, scaling, second order sections (SOS), signal to noise ratio (SNR), Simulink, uniform noise, wrapping

Contents

Notes

ACKNOWLEDGMENTS

We would like to gratefully acknowledge the support of Texas Instruments and The Mathworks in the development of this material.

PERMISSIONS

Figure 5.8 is adapted from [26]. Permission is pending.
Figures 5.14 through 5.19 are adapted from [23]. Permission is pending.

PRELIMINARY

This is a preliminary version of this book.

Wayne T. Padgett and David V. Anderson
September 2009

CHAPTER 1

Getting Started

More than 90 percent of signal-processing systems use finite-precision (fixed-point) arithmetic. This is because fixed–point hardware is lower cost and lower power, and often higher speed than floating–point hardware. The advantage of fixed–point hardware in a **digital signal processing** (DSP) **microprocessor** (μP) is largely due to the reduced data word size, since fixed–point is practical with 16 bit data, while floating–point usually requires 32 bit data. **Field programmable gate arrays** (FPGAs) gain similar advantages from fixed–point data word length and have the additional advantage of being customizable to virtually any desired wordlength.

Unfortunately, most academic coursework ignores the unique challenges of fixed–point algorithm design. This text is intended to fill that gap by providing a solid introduction to fixed–point algorithm design. You will see both the theory and the application of the theory to useful algorithms.

- Fixed-point signal processing is more difficult because it is non-linear

- Information tends to be harder to obtain (often scattered in journal articles, application notes, and lore)

- Fixed-point signal processing is usually not taught in universities.

- Matlab has a good toolbox for exploring fixed-point issues (we will discuss that in this text).

The notation used in this text is chosen to be familiar for those who have used the DSP First [18], Signal Processing First [19], or Discrete-Time Signal Processing [23] textbooks. We use these texts in our own teaching and have chosen definitions and topics of this text, especially in the introductory material, to be consistent with them.

1.1 FIXED-POINT DESIGN FLOW

One key principle to follow in fixed–point algorithm design is to minimize sources of bugs. To do this, develop the algorithm in an incremental way, only introduce a new layer of complexity when the first iteration is clearly working.

1. Develop a working algorithm in floating point.

2. Convert to fixed–point in a high level simulation tool. Use the simulation analysis tools to identify problems with limited precision and range. Resolve all precision related problems.

3. Implement fixed–point algorithm in real hardware and debug using comparison tests with high level simulation.

4. Iterate as necessary to resolve implementation limitations such as speed and limited resource problems.

1.2   TOOLS

There are tools available to aid in the development of fixed-point systems. None of these tools can automate the process sufficiently to absolve the designer of the need to understand fixed-point issues. Most of them are similar to woodworking tools—they make the process easier but they are only useful in the hands of someone who understands how to use them.

- **C and Assembly:** Most fixed–point programming for microprocessors is done in C and assembly. Assembly coding tends to be laborious, but worthwhile for "inner loops" where speed is critical. Analysis tools, plotting functions, etc. are not available in C.

- **Custom Tools:** It is possible to write custom fixed–point tools for carrying floating point and fixed–point values together and logging operations, etc. but they can be too slow to be useful.

- Matlab, **Filter Design Toolbox, Fixed Point Toolbox:** For script programming, Matlab provides a powerful environment for analysis, visualization, and debugging. The code examples in this text are given in Matlab.

  - Matlab can interface with libraries written in C.
  - Many people rewrite the essential parts of their code into C and continue to debug within the Matlab environment.
  - This provides a gradual path of transition.
  - An open source C++ library called IT++ has Matlab-like syntax and functions.
  - Using IT++ it is relatively easy to directly translate Matlab to C++.
  - When the C++ code operates as desired, gradual conversion to integer math is often the final step.

- **Simulink, Fixed Point Blockset:** A graphical programming approach has the advantage of imitating the signal flowgraphs we naturally use to describe algorithms, and it lends itself to parallel implementation in ways that script languages do not.

- **LabVIEW, Digital Filter Design Toolkit:** A graphical programming language with extensive support for data acquisition and instrument control. It also supports floating–point and integer arithmetic for signal processing with a large library of algorithms. Tools are available for fixed–point support and hardware targeting.

CHAPTER 2

DSP Concepts

This text is written with the assumption that you are familiar with basic DSP concepts such as would be covered in an undergraduate DSP course. However, to refresh these ideas and to establish notation, this chapter reviews a few key concepts.

2.1 BASIC SYSTEMS THEORY

2.1.1 LINEAR, TIME-INVARIANT SYSTEMS

Some systems are particularly difficult to understand and design. For example,

- Systems that change over time or that respond differently to the same input.

- Systems that respond to changes in input amplitude in "unnatural" ways

Restricting our study to **linear, time-invariant (LTI) systems** greatly simplifies analysis and design. It is a matter of debate whether there are any true LTI systems in the physical universe. However, many systems that are not LTI can be approximated that way. One notoriously *nonlinear* operation is quantization, a fundamental part of fixed-point systems. However, as we will see later, (nonlinear) quantization can be modeled as (linear) additive noise for most cases [28].

A system is **time-invariant** if delaying the input simply delays the output by the same amount of time.

A system is **linear** if it exhibits the following two properties

1. **Scaling**—scaling the input to the system causes the output to be scaled by the same amount.

2. **Superposition**—adding two inputs gives an output that is the same as the sum of the individual outputs.

Assuming an input signal, $x[n]$, and an output signal, $y[n]$, all systems whose functionality can be described in the form

$$y[n] = a_0 x[n] + a_1 x[n-1] + a_2 x[n-2] + \cdots$$

are **linear** and **time-invariant**. Furthermore, any LTI system can be completely characterized by its **impulse response**. The impulse function or signal is defined as

$$\delta[n] \;=\; \left\{ \begin{array}{ll} 1 & n = 0 \\ 0 & n \neq 0 \end{array} \right. \tag{2.1}$$

The impulse response, $h[n]$, is defined as the output of a system when the input, $x[n] = \delta[n]$, is an impulse. An LTI system may be labeled according to the nature of its impulse response—a system with a finite-length $h[n]$ is called an **FIR** (finite impulse response) system. A system having internal feedback may have an infinitely-long impulse response; in general, any system with an infinitely-long impulse response if referred to as an **IIR** system.

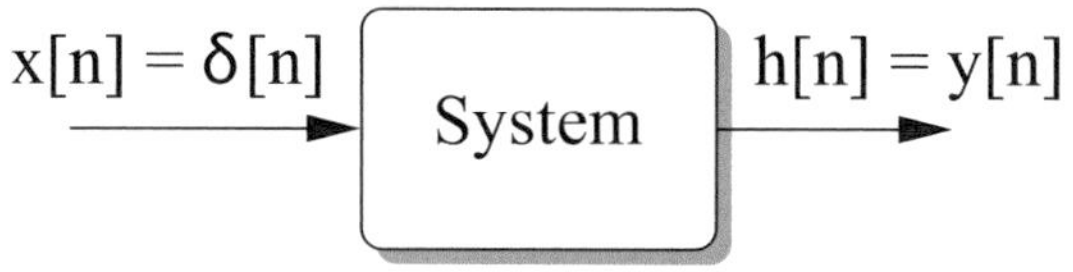

$$x[n] = \delta[n] \quad \boxed{\text{System}} \quad h[n] = y[n]$$

Figure 2.1: The impulse response $h[n]$ of a system is simply the output of the system when the input is an impulse.

2.1.2   DIFFERENCE EQUATIONS

A general useful class of discrete linear systems can be described by difference equations:[1]

$$y[n] = \sum_{k=1}^{N} a_k y[n-k] + \sum_{k=0}^{M} b_k x[n-k] \tag{2.2}$$

For this to be an LTI system it must be initially at rest, *i.e.* if $x[n] = 0$ for $n < n_0$, then $y[n] = 0$ for $n < n_0$.

The feed-back part of the system is given by $\sum_{k=1}^{N} a_k y[n-k]$ and the feed-forward part is given by $\sum_{k=0}^{M} b_k x[n-k]$. The parameters a_k, b_k, N, and M determine the characteristics of the system. *Filter design* involves establishing the system characteristics by selecting these parameters. If only the *feed-forward* part of equation (2.2) is present, then the system is FIR and we will see that the equation (2.2) reduces to a convolution equation.

2.1.3   CONVOLUTION

The convolution of the sequences $x[n]$ and $h[n]$ is defined as

$$y[n] = x[n] * h[n] = \sum_{k=-\infty}^{\infty} x[k]h[n-k] \tag{2.3}$$

Convolution defines the operation of a filter when $h[n]$ is the impulse response. That is, if we know the impulse response $h[n]$ of an LTI system, it's output $y[n]$ can be computed by convolving the input $x[n]$ with the impulse response.

---

[1]Actually, this is a causal linear system because the output is a function of only the current and prior inputs and outputs.

For FIR systems it is easy to relate the difference equation to the impulse response as $h[k] = b_k$. In this case, the operation of convolution is identical to computing the output using the difference equation. For IIR filters, there is no simple relationship between $h[k]$, b_k, and a_k but convolution gives the same result as computing using the difference equation.

Some important properties of convolution are given below.

commutativity: $$x[n] * h[n] = h[n] * x[n] \tag{2.4}$$
associativity: $$(x[n] * h_1[n]) * h_2[n] = x[n] * (h_1[n] * h_1[n]) \tag{2.5}$$
linearity: $$x[n] * (ah_1[n] + bh_2[n]) = ax[n] * h_1[n] + bx[n] * h_2[n] \tag{2.6}$$
identity: $$x[n] * \delta[n] = x[n] \tag{2.7}$$
shift: $$x[n] * \delta[n - n_0] = x[n - n_0] \tag{2.8}$$

Note, the commutativity rule implies

$$y[n] = \sum_{k=-\infty}^{\infty} x[n-k]h[k] = \sum_{k=-\infty}^{\infty} x[k]h[n-k]. \tag{2.9}$$

A few important things to note about convolution:

- Any LTI system can be described using an impulse response and the output can be computed using convolution.

- If the input is of length N and the impulse response is of length M, then the output is of length $M + N - 1$.

- The total multiplies required to perform convolution on an entire signal is $N * M$.

2.2   Z-TRANSFORM DOMAIN SYSTEMS ANALYSIS

The z-transform is a critically important tool for analyzing filters and the associated finite precision effects[2]

Definition 2.1.   The z-transform of $x[n]$ is defined by

$$X(z) = \sum_{n=-\infty}^{\infty} x[n]z^{-n}. \tag{2.10}$$

Example 2.2.1 (z-transform of signal).

$$x[n] = \begin{bmatrix} 1 & 2 & 3 \end{bmatrix} \quad \text{gives} \quad X(z) = 1 + 2z^{-1} + 3z^{-2}$$

[2]We will use the term *filter* and the term *LTI system* interchangeably from this point forward.

Example 2.2.2 (z-transform of impulse response).

$$h[n] = \begin{bmatrix} 1 & -1 & 2 \end{bmatrix} \quad \text{gives} \quad H(z) = 1 - z^{-1} + 2z^{-2}$$

The z-transform can also be found for the difference equation representation of a signal. This operation can be reduced the following simple steps:

- Wherever you see $[n - n_0]$ put a z^{-n_0}.

- Replace all little letters ($x[n]$, $y[n]$) with big letters ($X(z)$, $Y(z)$).

Then

$$y[n] = \sum_{k=1}^{N} a_k y[n - k] + \sum_{k=0}^{M} b_k x[n - k]$$

becomes

$$Y(z) = \sum_{k=1}^{N} a_k Y(z) z^{-k} + \sum_{k=0}^{M} b_k X(z) z^{-k}.$$

By collecting terms we have

$$Y(z) \left(1 - \sum_{k=1}^{N} a_k z^{-k} \right) = X(z) \left(\sum_{k=0}^{M} b_k z^{-k} \right).$$

Alternatively we can write

$$Y(z) = \frac{\sum_{k=0}^{M} b_k z^{-k}}{1 - \sum_{k=1}^{N} a_k z^{-k}} X(z). \tag{2.11}$$

One very useful property of the z-transform is that it converts time-domain convolution into frequency-domain multiplication, so that when

$$y[n] = x[n] * h[n], \qquad Y(z) = X(z)H(z). \tag{2.12}$$

Relating this back to the z-transform of difference equations from above implies

$$H(z) = \frac{\sum_{k=0}^{M} b_k z^{-k}}{1 - \sum_{k=1}^{N} a_k z^{-k}}.$$

$H(z)$ is called the transfer function. The transfer function of the difference equation is often written as

$$H(z) = \frac{\sum_{k=0}^{M} b_k z^{-k}}{1 - \sum_{k=1}^{N} a_k z^{-k}} = \frac{B(z)}{A(z)} \tag{2.13}$$

where $A(z) = 1 - \sum_{k=1}^{N} a_k z^{-k}$ and $B(z) = \sum_{k=0}^{M} b_k z^{-k}$ are polynomials in z^{-1}.

2.2.1 POLES AND ZEROS

Definition 2.2. The poles and zeros of a transfer function are defined as the roots of the denominator and numerator polynomials, respectively.

Note, to keep the algebra clear, it is usually best to work in positive powers of z. For example, $H(z) = 1 - z^{-1} + 2z^{-2}$ may be written as

$$H(z) = 1 - z^{-1} + 2z^{-2}\left(\frac{z^2}{z^2}\right) = \frac{z^2 - z + 2}{z^2}$$

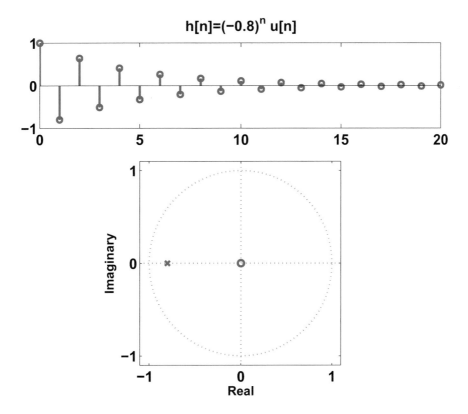

Figure 2.2: The impulse and pole-zero plot for a single pole IIR system. Poles are indicated by 'x' and zeros by 'o'.

Example 2.2.3 (poles and zeros). Given the system

$$y[n] = ay[n-1] + x[n], \tag{2.14}$$

the impulse response is

$$h[n] = a^n u[n], \tag{2.15}$$

and the z-transform is

$$H(z) = \frac{1}{1 - az^{-1}} = \frac{z}{z - a} \tag{2.16}$$

This system has a single pole at $z = a$ and a zero at $z = 0$. Figure 2.2 shows the impulse response, $h[n]$, as well as a plot of the pole and zero on the complex plane for $a = -0.8$.

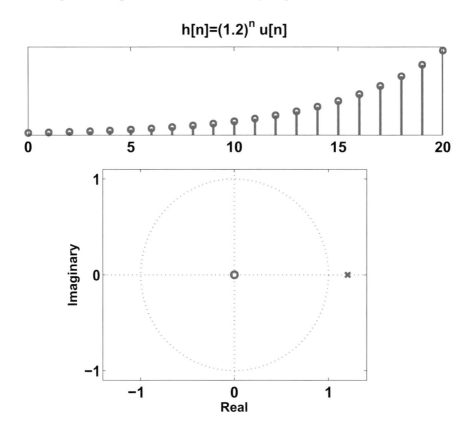

Figure 2.3: The impulse and pole-zero plot for a single pole *unstable* IIR system.

When $a = 1.2$ the impulse response grows rather than dying away. The result is that a small input can yield an infinite output and; therefore, the system is *unstable*. Note that (coincidentally?) the magnitude of the pole is greater than 1.

In general, poles with a magnitude greater than 1 are associated with unstable systems. For this reason (and others), a circle of radius 1 (called the *unit circle*) is usually superimposed on pole-zero plots as in Figs. 2.2 & 2.3. Poles outside the unit circle indicate an unstable system. Poles on the unit circle are technically unstable also, but they have impulse responses that rings or oscillates instead of blowing up. Zeros can be inside or outside or on the unit circle without ill effect.

2.3   BLOCK DIAGRAMS AND FILTER IMPLEMENTATION

Block diagrams are a useful way of representing systems and, in particular, the implementation details. The basic building elements are adders, multipliers, delays, and subsystems.

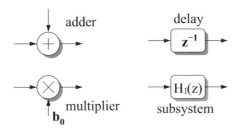

Figure 2.4: Block diagram building blocks.

The *direct form I* filter representation (Fig. 2.5) is so called because it is the direct implementation of the difference equation. The FIR or feed-forward portion of the system is first and the feed-back portion of the system is second; just as in the equation.

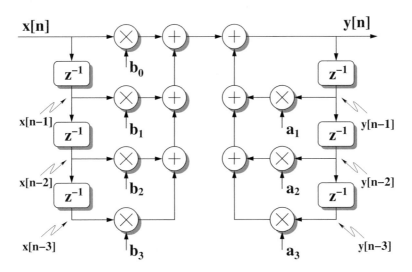

Figure 2.5: Direct Form I implementation of a filter. This is a direct implementation of the difference equation.

Since convolution is commutable, it is possible to change the order of the two parts of the filter without affecting the response.[3] In this form, (Fig. 2.6) the feed-back operation is performed first, followed by the FIR portion. A significant advantage of this form is that the same value is used as the input to both delay lines. Therefore, one of the delay lines may be eliminated, saving memory, as shown in Fig. 2.7. The resulting form is called *direct form II*.

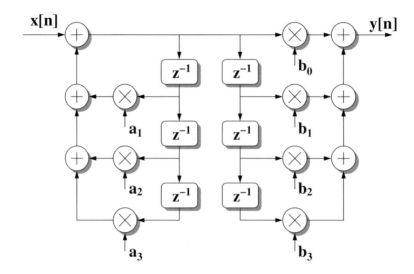

Figure 2.6: Linearity makes it possible to swap the order of the feed-forward and feed-back portions of the filter without changing the filter function.

There are many equivalent filter forms but most are not covered in basic signal processing courses because they are equivalent. However, fixed-point systems do not operate in an ideal manner and so these different forms are no longer equivalent. In some cases, one filter form will perform much better than another. Therefore, we will introduce several more filter forms.

2.3.1   TRANSPOSE FILTERS

The transpose form of a filter is generated from a block diagram using the following steps

1. Reverse the direction of each arrow

2. Each adder becomes a branch

3. Each branch becomes an adder

4. Swap the input and output

[3]This assumes ideal computation; in practice, the output may be different for the two forms as a result of finite-precision arithmetic as will be seen later.

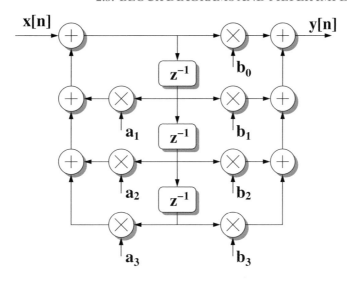

Figure 2.7: Combining the delay lines in Fig. 2.6 yields Direct Form II. This form takes $\frac{1}{2}$ the memory of DFI.

The resulting filter is equivalent to its non-transpose form.

The direct form II transpose filter is shown in Fig. 2.8. For hardware systems, there are several advantages to this form. Namely, the same values are at the input of all multipliers, saving register load times and allowing implementations such as shared sub-expressions. Also, the adders are followed by delays, so latching the adder output naturally implements the delay without the need for a separate register.

2.3.2 CASCADED SECOND-ORDER SECTIONS

There are a variety of filter structures, each with its own set of advantages and disadvantages. This following list is an incomplete but representative list of other filter structures.

- FIR Lattice filters

- Folded FIR filters

- IIR Lattice filters (and transpose structures)

- Polyphase FIR structures

- Parallel IIR structures

- All-pass structures

- Composite all-pass structure filters

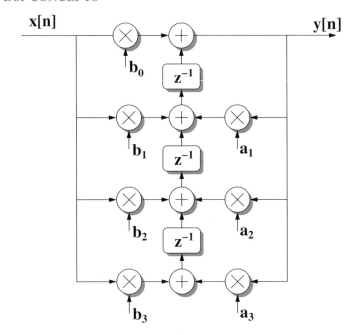

Figure 2.8: The Direct Form II transpose form of a filter.

- Gray-Markel structure

One structure that is of particular importance is the cascaded second-order section (SOS) structure. This structure will be discussed in detail in later chapters but is shown here in Fig. 2.9.

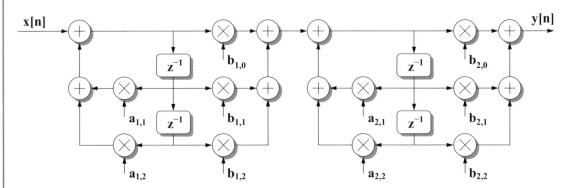

Figure 2.9: An IIR filter composed of cascaded 2nd order filters. Any direct form filter can be converted to this form.

2.4 FREQUENCY RESPONSE

Physical signals can be decomposed into frequency components and the effect of an LTI system on those signals may be understood by understanding the effect of the LTI system on each of the components. Such analysis lends intuitive understanding of the effect of filters on different types of signals.

The complex exponential and Euler's identity form the foundation for frequency analysis of signals. Euler's identity relates complex exponentials to sine and cosines as

$$e^{j\alpha} = \cos\alpha + j\sin\alpha \tag{2.17}$$

where $j = \sqrt{-1}$ (the unit imaginary number). Euler's identity allows us to simplify trigonometric operations by replacing them with operations on exponents. This forms the basis of the FFT (Fast Fourier Transform) that will be discussed later. This also forms the basis for analyzing filters and signals.

After a little manipulation, Euler's identity can be written as:

$$\begin{aligned}
\cos\alpha &= \frac{1}{2}\left(e^{j\alpha} + e^{-j\alpha}\right) \\
\sin\alpha &= \frac{1}{2j}\left(e^{j\alpha} - e^{-j\alpha}\right)
\end{aligned} \tag{2.18}$$

We can use Euler's identity in a different way to express our sinusoid function

$$\begin{aligned}
A\cos(\omega t + \phi) &= \Re\left\{Ae^{j(\omega t + \phi)}\right\} \\
&= \Re\left\{Ae^{j(\phi)}e^{j(\omega t)}\right\} \\
&= \Re\left\{Xe^{j(\omega t)}\right\}
\end{aligned} \tag{2.19}$$

where $X = Ae^{j(\phi)}$ is a complex number.

Another concept that plays an important role in evaluating the filter responses is the *unit circle*. The unit circle is a circle with a radius of one, centered at the origin of the complex plane (see Fig. 2.10). It is the line that is described by plotting the imaginary parts of $z = e^{-j\theta}$ vs. the real part as θ ranges between $0 \leq \theta \leq 2\pi$. Note that $z = e^{-j\theta}$ is periodic with $z = e^{j\theta} = e^{j\theta + 2\pi k}$ for integer values of k.

Some key points on the unit circle are summarized below.

$$\begin{aligned}
\theta = 0 &\implies e^{j0} = 1 \\
\theta = 2\pi k &\implies e^{j2\pi k} = 1 \\
\theta = \frac{\pi}{2} &\implies e^{j\frac{\pi}{2}} = j \\
\theta = \pi &\implies e^{j\pi} = -1 \\
\theta = \frac{-\pi}{2} &\implies e^{-j\frac{\pi}{2}} = -j
\end{aligned} \tag{2.20}$$

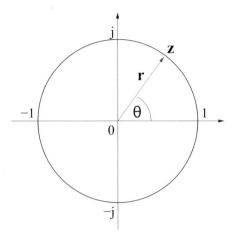

Figure 2.10: All points on the unit circle are defined by an angle θ.

2.4.1   FREQUENCY RESPONSE FROM THE Z-TRANSFORM

Definition 2.3.   The frequency response of an LTI system is defined as

$$H(e^{j\omega}) = H(z)|_{z=e^{j\omega}} = \sum_{n=-\infty}^{\infty} h[n]e^{-j\omega n} \tag{2.21}$$

Frequency response can be seen as evaluating the z-transform along the unit circle in the z plane.

When zeros are near the unit circle, $|H(e^{j\omega})|$ tends to drop near the zero. When poles are near the unit circle, $|H(e^{j\omega})|$ tends to increase near the pole. When poles are near the unit circle they cause very high filter gains, which are difficult to handle in fixed-point.

2.4.2   FREQUENCY RESPONSE EXAMPLES
Figures 2.11-2.16 illustrate several different filters, showing the impulse response, pole-zero plot, and frequency response for each.

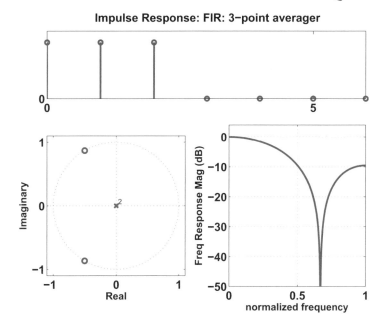

Figure 2.11: 3-point running averager

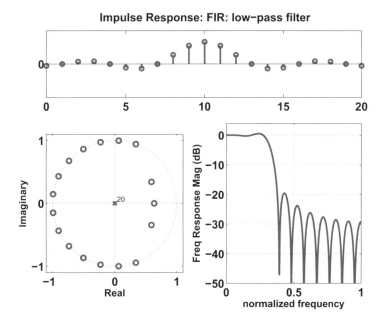

Figure 2.12: Low-pass FIR filter (21st order)

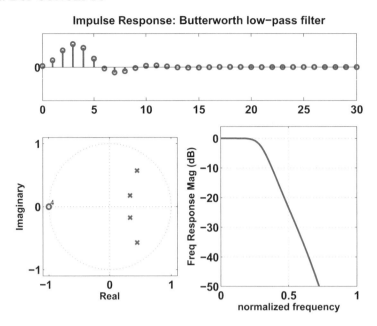

Figure 2.13: Low-pass Butterworth filter (4th order)

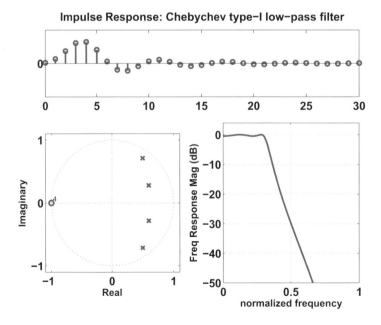

Figure 2.14: Low-pass Chebyshev type-I filter (4th order)

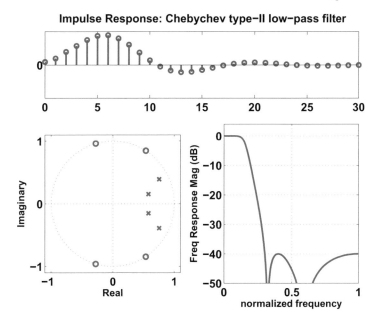

Figure 2.15: Low-pass Chebyshev type-II filter (4th order)

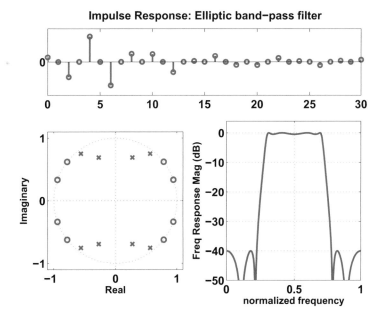

Figure 2.16: Band-pass elliptic (5th order)

CHAPTER 3

Random Processes and Noise

Nearly all signals are random! Noise is the first thing that comes to mind when one thinks of random signals but even signals which have some structure such as speech or images are essentially random signals with some structure. For these reasons alone, an essential part of signal processing is the analysis and modeling of random signals. Finite-precision signal processing adds another important reason—quantization and rounding are not LTI operations and a general analysis of these operations is intractable. However, by casting the problem in terms of random signals, much useful understanding may be gained relative to quantization and rounding.

In order to analyze random signals we must first define concepts and characteristics of random numbers. In practice, it is possible to use many of the same analysis techniques that are used for deterministic signals. Finally, with random signals, we are concerned with statistical outcomes.

3.1 RANDOM VARIABLES

Definition 3.1. A **random variable** is a variable which represents the outcome of a random event so that the possible outcomes can be handled numerically in equations.

Each time the random variable is inspected, it may return a different value. Since the random value cannot be known in advance, it is often denoted symbolically. The information available about the likelihood of each possible value or range of values is expressed in cumulative density functions.

A **random signal** is composed of values generated by a random variable.

Definition 3.2. The cumulative density (or distribution) function (CDF), P_x of a random variable x is defined as the probability that x is less than or equal to some value u.

$$P_x(u) = P(x \leq u) \tag{3.1}$$

Note, $P_x(-\infty) = 0$ and $P_x(\infty) = 1$.

Definition 3.3. The probability density function (PDF) is the derivative of the CDF.

$$p_x(u) = \frac{d P_x(u)}{du} \tag{3.2}$$

alternatively,

$$P_x(u) = \int_{-\infty}^{u} p_x(v)dv \tag{3.3}$$

This is commonly used and intuitively: *if $p_x(u_1) > p_x(u_2)$ then x is more likely to take on the value of u_1 than u_2*.[1]

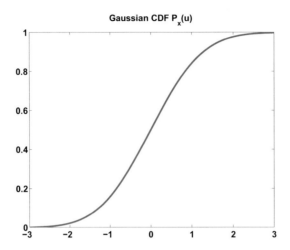

Figure 3.1: Gaussian cumulative density function (CDF)

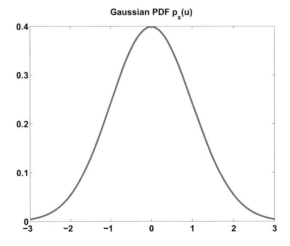

Figure 3.2: Gaussian probability density function (pDF)

---

[1] Since the probability of a continuous random variable taking on an exact value is always zero, this intuitive statement should really be thought of as "*x* is more likely to take on a value *near u_1* than u_2," but this more precise statement seems less memorable.

Multiple Random Variable Functions   The PDF can also be defined for two random variables

$$P_{xy}(u, v) = P(x \leq u, y \leq v) \tag{3.4}$$

and

$$p_{xy}(u, v) = \frac{d^2 P_{xy}(u, v)}{du\,dv} \tag{3.5}$$

marginal densities

$$p_x(u) = \int_{-\infty}^{\infty} p_{xy}(u, v)\,dv \tag{3.6}$$

$$p_y(v) = \int_{-\infty}^{\infty} p_{xy}(u, v)\,du. \tag{3.7}$$

Definition 3.4.   Statistical Independence: Two random variables x and y are statistically independent if and only if their joint PDF factors as

$$P_{xy}(u, v) = P_x(u) P_y(v) \tag{3.8}$$

Intuitively, samples of a random process are independent if knowing one sample imparts no information about the other.

3.1.1   EXPECTATIONS AND MOMENTS

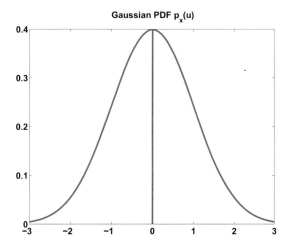

Figure 3.3: The mean of a Gaussian corresponds to the peak of its PDF.

The mean or average is often considered the first or basic statistic describing a random quantity. The mean is the expected value and is defined as

$$\mu_x = E\{x\} = \int_{-\infty}^{\infty} x p_x(x) dx \tag{3.9}$$

The expected value or mean has several useful properties that arise from the fact that it is found using a linear operation.

$$E\{x + y\} = E\{x\} + E\{y\} \tag{3.10}$$
$$E\{ax\} = aE\{x\}. \tag{3.11}$$

Correlation   The cross-correlation of two random variables gives an indication of the linear relationship between them. The cross-correlation of two random variables x and y is defined as

$$\phi_{xy} = E\{xy\} = \int_{-\infty}^{\infty} uv p_{xy}(u, v) du dv \tag{3.12}$$

For *statistically independent* or *uncorrelated* random variables

$$E\{xy\} = E\{x\} \cdot E\{y\} \tag{3.13}$$

- Independence implies uncorrelatedness

- Uncorrelatedness does **not** imply independence

Variance   Variance is a measure of how much a random variable may deviate from its mean. The variance is defined as

$$\sigma_x^2 = var[x] = E\left\{|x - \mu_x|^2\right\} \tag{3.14}$$

By multiplying out the square it can be shown

$$\sigma_x^2 = var[x] = E\left\{|x|^2\right\} - |\mu_x|^2 \tag{3.15}$$

With random signals having a mean of zero (also called zero-mean signals), the variance corresponds to the power of the signal.

Covariance   The covariance is defined as

$$\gamma_{xy} = E\left\{(x - \mu_x)(y - \mu_y)\right\} = \phi_{xy} - \mu_x \mu_y \tag{3.16}$$

This could be thought of as the "prediction power with the mean removed."

Autocorrelation   The autocorrelation is defined as

$$\phi_{xx}[n, m] = E\left\{x[n]x^*[m]\right\} \tag{3.17}$$

and it provides information about how correlated various samples in a signal are. Note, $x^*[m]$ is the complex-conjugate of $x[m]$.

3.1.2   STATIONARY AND ERGODIC PROCESSES

Definition 3.5.   Random Process: A sequence of random variables $x[n]$ indexed by an integer n is a discrete-time random process (or random signal).

Definition 3.6.   Stationary: If the statistics do not change as a function of n, then $x[n]$ is **stationary**.

For stationary signals,

- The mean is constant for the entire duration of the signal.

- The autocorrelation is a function only of the index shift. That is $\phi_{xx}[n + m, n]$ can be written simply as

$$\phi_{xx}[m] = E\left\{x[n + m]x^*[n]\right\} \tag{3.18}$$

Some useful signals are strictly stationary. A less rigid type of stationarity is called *wide-sense stationarity*. A signal is wide-sense stationary if its *mean* and *autocorrelation* functions do not change with time.

A practical problem with random systems is calculating or estimating the mean and variance. Since the underlying PDF is almost never known, the mean and variance must be estimated using other methods. For an *ergodic* process, the true underlying mean and variance can be approximated by averaging samples over time. This is extremely useful (and it works for anything we might be worried about). Ergodicity implies that the following estimates approach the true values for large values of L:

$$\hat{\mu}_x = \frac{1}{L}\sum_{n=0}^{L-1}x[n] \qquad \hat{\sigma}_x^2 = \frac{1}{L}\sum_{n=0}^{L-1}|x[n] - \hat{\mu}_x|^2 \tag{3.19}$$

$$\hat{\phi}_{xx}[m] = \frac{1}{L}\sum_{n=0}^{L-1}x[n + m]x^*[n] \tag{3.20}$$

The estimate of the variance above is called the sample variance and it actually yields a biased estimate of the true variance. An unbiased estimate of the variance is found using

$$\hat{\sigma}_x^2 = \frac{1}{L - 1}\sum_{n=0}^{L-1}|x[n] - \hat{\mu}_x|^2 \tag{3.21}$$

Matlab functions exist for these operations.

- `mean` — compute the sample mean ($\hat{\mu}_x$)

- `var(x,1)` — compute the sample variance ($\hat{\sigma}_x^2$)

- `var(x)` — compute the *unbiased* variance estimate ($\hat{\sigma}_x^2$)

- `xcorr(x,'biased')` — compute the autocorrelation ($\hat{\phi}_{xx}[m]$)

  - For the `xcorr` function, the biased estimate is smaller at the edges because it is using fewer samples to estimate it.

  - The unbiased estimate is normalized by the number of samples used to estimate each sample–this makes the end points noisier.

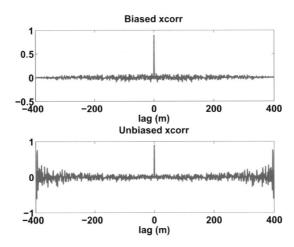

Figure 3.4: Cross-correlation of a white noise sample.

3.1.3   DEFINITIONS AND PROPERTIES

Assuming stationary processes:

$$\text{autocorrelation} \quad \phi_{xx}[m] = E\left\{x[n+m]x^*[n]\right\} \tag{3.22}$$

$$\text{autocovariance} \quad \gamma_{xx}[m] = E\left\{(x[n+m]-\mu_x)(x[n]-\mu_x)^*\right\} \tag{3.23}$$

$$\text{crosscorrelation} \quad \phi_{xy}[m] = E\left\{x[n+m]y^*[n]\right\} \tag{3.24}$$

$$\text{crosscovariance} \quad \gamma_{xy}[m] = E\left\{(x[n+m]-\mu_x)(y[n]-\mu_y)^*\right\} \tag{3.25}$$

Properties: Covariance

$$\gamma_{xx}[m] = \phi_{xx}[m] - |\mu_x|^2 \tag{3.26}$$

$$\gamma_{xy}[m] = \phi_{xy}[m] - \mu_x \mu_y^* \tag{3.27}$$

Autocorrelation and autocovariance only differ by the mean. For zero-mean processes, they are the same.

Properties: Zero Shift

$$\phi_{xx}[0] = E\left\{|x[n]|^2\right\} \qquad \text{mean-square} \tag{3.28}$$

$$\gamma_{xx}[0] = \sigma^2 \qquad \text{variance} \tag{3.29}$$

$$|\phi_{xx}[m]| \le \phi_{xx}[0] \tag{3.30}$$

$$|\gamma_{xx}[m]| \le \gamma_{xx}[0] \tag{3.31}$$

The peak of the autocorrelation and autocovariance functions is always at zero. This essentially means that nothing is more useful for predicting nearby values of a sequence than the value of itself (zero shift).

Properties: Autocorrelation Functions   Most random processes become less correlated with greater time shifts (larger values of *m*). Hence, the autocorrelation functions die out and have finite energy. Therefore, autocorrelation functions have Fourier and *z*-transforms. Note, this is not true with stationary random sequences since they must go on forever and subsequently have infinite energy.

3.2   RANDOM PROCESSES AND FOURIER ANALYSIS

3.2.1   FOURIER TRANSFORM OF CORRELATION AND COVARIANCE

We can make use of Fourier transforms of the correlation and covariance functions:

$$\Phi_{xx}(e^{j\omega}) = \mathcal{F}\{\phi_{xx}[m]\} = \sum_{k=-\infty}^{\infty} \phi_{xx}[k]e^{-j\omega k} \tag{3.32}$$

$$\Gamma_{xx}(e^{j\omega}) = \mathcal{F}\{\gamma_{xx}[m]\} = \sum_{k=-\infty}^{\infty} \gamma_{xx}[k]e^{-j\omega k} \tag{3.33}$$

Note,

$$\Phi_{xx}(e^{j\omega}) = \Gamma_{xx}(e^{j\omega}) + 2\pi |\mu_x|^2 \delta(\omega), \quad |\omega| < \pi \tag{3.34}$$

Similarly, the inverse transforms are defined as:

$$\gamma_{xx}[m] = \frac{1}{2\pi} \int_{-\pi}^{\pi} \Gamma_{xx}\left(e^{j\omega}\right) e^{j\omega m} d\omega \tag{3.35}$$

$$\phi_{xx}[m] = \frac{1}{2\pi} \int_{-\pi}^{\pi} \Phi_{xx}\left(e^{j\omega}\right) e^{j\omega m} d\omega \tag{3.36}$$

One useful relationship stemming from these definitions is Parseval's Theorem:

$$E\left\{|x[n]|^2\right\} = \phi_{xx}[0] = \sigma_x^2 + |\mu_x|^2 \tag{3.37}$$

$$= \frac{1}{2\pi} \int_{-\pi}^{\pi} \Phi_{xx}\left(e^{j\omega}\right) d\omega \tag{3.38}$$

$$\sigma_x^2 = \gamma[0] = \frac{1}{2\pi} \int_{-\pi}^{\pi} \Gamma_{xx}\left(e^{j\omega}\right) d\omega \tag{3.39}$$

3.2.2   POWER SPECTRAL DENSITY

Noise signals are *power signals*, not *energy signals*. The Fourier Transform of a power signal does not converge because, as mentioned above, it has infinite energy. However, it is still possible to talk about the frequency distribution of the noise power. In this case, the frequency distribution depends on how "related" nearby samples are. If nearby samples are very similar, the process is slowly varying and has more power at low frequencies. If nearby samples are "unrelated," the process changes rapidly and has more power at high frequencies.

The relationship of nearby samples can be expressed as autocorrelation. Recall that σ_x^2 is the power in a zero-mean random process. We can define the power spectrum $P_{xx}(\omega) = \Phi_{xx}\left(e^{j\omega}\right)$. The quantity

$$\frac{1}{2\pi} \int_{\omega_1}^{\omega_2} P_{xx}(\omega) d\omega \tag{3.40}$$

represents the power in frequencies between ω_1 and ω_2. Integrating over all frequencies gives the total energy

$$E\left\{|x[n]|^2\right\} = \frac{1}{2\pi} \int_{-\pi}^{\pi} P_{xx}(\omega) d\omega. \tag{3.41}$$

White noise (in discrete-time) is noise in which each sample is uncorrelated with every other sample. White noise gets its name from its power spectral density (PSD). The PSD of white noise is constant, similar to white light which contains all optical wavelengths, with

$$P_{xx}(\omega) = \sigma_x^2 \qquad \text{(with zero mean)} \tag{3.42}$$

Note also that

$$\sigma_x^2 = \frac{1}{2\pi} \int_{-\pi}^{\pi} \sigma_x^2 d\omega \tag{3.43}$$

and

$$\phi_{xx}[m] = \sigma_x^2 \delta[m], \qquad \Phi_{xx}\left(e^{j\omega}\right) = \sigma_x^2 \tag{3.44}$$

3.2.3   FILTERING A RANDOM SEQUENCE
Suppose

$$y[n] = h[n] * x[n] \tag{3.45}$$

where $x[n]$ represents a random process and $h[n]$ is a real-valued impulse response. Then

$$\Phi_{yy}\left(e^{j\omega}\right) = \left|H\left(e^{j\omega}\right)\right|^2 \Phi_{xx}\left(e^{j\omega}\right) \tag{3.46}$$

and

$$\phi_{yy} = (h[m] * h[-m]) * \phi_{xx}[m] \tag{3.47}$$

Note, for real $h[n]$, $\mathcal{F}\{h[m] * h[-m]\} = \left|H\left(e^{j\omega}\right)\right|^2$

Example 3.2.1 (filtering and autocorrelation). Suppose $y[n]$ is the output of a filter. Use the relation

$$\phi_{yy}[m] = (h[m] * h[-m]) * \phi_{xx}[m]$$

to determine $\phi_{yy}[m]$ when $h[m] = [\overset{m=0}{1} \quad 2 \quad -3]$ and $\phi_{xx}[m] = [4 \quad \overset{m=0}{5} \quad 4]$. Note that by the definitions given above, $\phi_{xx}[m]$ and $\phi_{yy}[m]$ must be even, and $\sigma_x^2 = 5$. Consider using the conv() function in Matlab.

   You should get

$$\phi_{yy}[m] = [-12 \ -31 \ 24 \ \overset{m=0}{38} \quad 24 \ -31 \ -12].$$

Recall $\sigma_y^2 = E\{y[k]y[k]\} - \mu_y$. Since $\mu_y = 0$,

$$E\{y[k]y[k]\} = \phi_{yy}[0] = \sigma_y^2 = 38.$$

3.2.4   FILTERING A WHITE RANDOM SEQUENCE
If the input, $x[n]$, to a system is white noise then

$$\Phi_{xx}\left(e^{j\omega}\right) = \sigma_x^2. \tag{3.48}$$

Then if the system is described by a transfer function $H(z)$

$$\Phi_{yy}\left(e^{j\omega}\right) = \left|H\left(e^{j\omega}\right)\right|^2 \Phi_{xx}\left(e^{j\omega}\right). \tag{3.49}$$

This can be expressed in the time domain as

$$\sigma_y^2 = \sigma_x^2 \sum_n |h[n]|^2 . \tag{3.50}$$

Exercise 3.2.2 (filtering white noise). Suppose

$$y[n] = h[n] * x[n]$$

where $x[n]$ represents a zero mean random process. Use the relation

$$\sigma_y^2 = \sigma_x^2 \sum_n |h[n]|^2$$

determine σ_y^2 when $h[m] = [\overset{m=0}{1}\ 2\ -3]$ and $\phi_{xx}[m] = 5\delta[m]$. Note that in the previous example, the random process $x[n]$ was non-white, but the identity to be used here requires white noise input to the filter. How can you tell if $x[n]$ is a white process or not? [2]

3.2.5   PERIODOGRAMS

Before discussing periodograms, it is important to briefly discuss windowing. Any given random sequence, $x[n]$ can be assumed to be infinite in length. Thus, analyzing any finite subsequence is equivalent to multiplying the subsequence by 1's and the rest of the signal by 0 prior to analysis. This function of, in this case, 1's and 0's is called a rectangular window (Fig. 3.5). There are also many other kinds of windows that can be used to achieve different effects.

The effect of a window on the sequence being evaluated can be understood by recalling that convolution in the time domain corresponds to multiplication in the frequency domain. And multiplication in the time domain corresponds to convolution in the frequency domain. Convolution in frequency spreads or blurs the spectrum according to the shape of the transform of the window. There are many windows available, each with different trade-offs. The primary trade-offs are:

- Mainlobe width

- Highest sidelobe

- Sidelobe roll-off

Windows play an important role in estimating the PSD of a signal.

Given a sequence $x[n]$, one straight-forward estimate [23] of $P_{xx}(\omega)$ is given by

$$V\left(e^{j\omega}\right) = \sum_{n=0}^{L-1} w[n]x[n]e^{-j\omega n} \tag{3.51}$$

---

[2] $\sigma_y^2 = 70$. A white process has $\phi_{xx}[m] = 0$ for all $m \neq 0$.

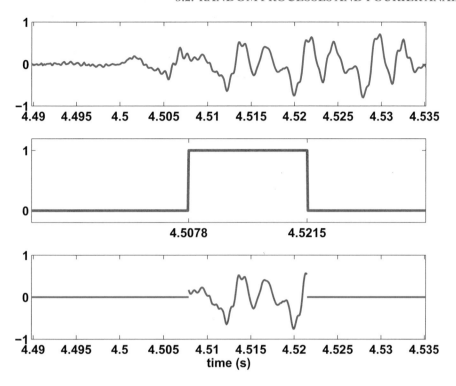

Figure 3.5: A rectangular time window applied to a speech signal.

where $w[n]$ is a window function (can be all 1's) and $V\left(e^{j\omega}\right)$ is the DTFT of $w[n]x[n]$. Then

$$P_{xx}(\omega) \approx \frac{1}{LU} \left| V\left(e^{j\omega}\right) \right|^2 = I(\omega) \tag{3.52}$$

where $U = \frac{1}{L} \sum_{n=0}^{L-1} w^2[n]$ to normalize with respect to the window power.

There are two problems with this simple estimate of the PSD. First, the window function distorts the spectrum as discussed above. Namely, while the window isolates a section of the signal to analyze, The resulting spectrum estimate is the true spectrum estimate convolved with the frequency response of the window. Second, the variance of the estimate does not improve with longer windows. This is particularly troubling because it implies that one cannot improve the estimate of the PSD by using more samples.

The solution to the first problem is to use longer windows and/or select window functions with more desirable characteristics. The solution to the second problem is the *periodogram*.

To improve the variance of the PSD estimate, multiple estimates of the PSD are found using different portions of the signal. These PSD estimates are then averaged together to form the periodogram. The variance of the periodogram estimate is decreased relative to the naive method by a factor of $\frac{1}{K}$ where K is the number of estimates in the average.

A further improvement to the periodogram is called the Welch method. With this method, overlapping blocks of length L are selected with an overlap of $L/2$. This reduces the variance further by $\frac{1}{2}$ since twice as many estimates may be performed on a given data sequence.

CHAPTER 4

Fixed Point Numbers

4.1 BINARY ARITHMETIC

While DSP algorithms are usually developed using floating–point arithmetic, most computer computation is accomplished using adders and multipliers designed for integer operations on fixed-length binary words.

The description of binary arithmetic given here is not intended to represent a complete presentation of hardware optimized algorithms, although some will be discussed in a later chapter, nor is it intended to be a theoretical discussion of number systems. Our intent here is really to review. We want to remind the reader of enough binary arithmetic to enable hand verification of the concepts discussed — mainly to enhance intuition and visualization. Excellent references on hardware algorithms can be found in Hennessy and Patterson [7] and Weste and Harris [27], and a good discussion of number systems can be found in Knuth [15].

4.1.1 UNSIGNED BINARY REPRESENTATION

The definition of binary numbers that is familiar to most people is the unsigned integer form given by

$$x_{(10)} = \sum_{i=0}^{n-1} b_i 2^i. \tag{4.1}$$

The notation here is defined as

b_i binary digit value,
n number of integer digits,
$x_{(10)}$ the base-10 (decimal) number.

The same idea can be extended to base 8 (octal), 16 (hexadecimal), etc. The base value is known as the *radix*. For general radix formulation, we can also represent fractions just as we do in decimal, with digits to the right of the decimal point — now called the *radix point*.

For the more general case, we have

$$x_{(10)} = \sum_{i=-m}^{n-1} b_i k^i. \tag{4.2}$$

with the notation of

b_i digit value

k	radix (for binary, this is 2)
n	digits to the left of the radix point
m	digits to the right of the radix point
M	total number of digits (bits)
$x_{(10)}$	the base-10 (decimal) number

Converting Binary to Decimal   If we assume a radix point to the right of the last digit (as we would do in decimal), the binary number 00010101 is interpreted as

$$00010101_{(2)} = 0 \cdot 2^7 + 0 \cdot 2^6 + 0 \cdot 2^5 + 1 \cdot 2^4 + 0 \cdot 2^3 + 1 \cdot 2^2 + 0 \cdot 2^1 + 1 \cdot 2^0 = 21_{(10)}.$$

Similarly, the binary number 0.10 is interpreted as

$$0.10_{(2)} = 0 \cdot 2^0 + 1 \cdot 2^{-1} + 0 \cdot 2^{-2} = 0.5_{(10)}.$$

Exercise 4.1.1 (binary to decimal). Try converting the binary values 011.011 and 0110.1010 to decimal for yourself.[1]

Converting Decimal to Binary   The standard algorithm for converting decimal numbers to binary numbers is to divide by two, taking the integer part of the quotient for the next stage, and the remainder as the first bit in the converted binary number. The process repeats until the quotient is zero.

One way to see why this procedure works (and thus more easily remember it), is to consider the number written in binary. Each time we divide by two and take the remainder, we are determining if the number is even or odd (remainder of 1 for odd numbers, 0 for even numbers). Clearly, even binary numbers will have a least significant bit (LSB) of 0, and odd binary numbers will have an LSB of 1. Therefore, the remainder will also be the LSB of the number under consideration. Since the integer quotient from the previous stage is equivalent to the integer result of right shifting the binary number, determining whether it is even or odd determines the value of the next bit, and so on. An example calculation is shown in Figure 4.1 where you can see that the right shifted binary values are the same as the integer quotients.

A similar procedure works for converting decimal fractions to binary fractions, and for similar reasons. If you take a decimal fraction and multiply by 2, the resulting integer part will be zero or one. It will also represent the first fractional binary digit to the right of the radix point. Discard the integer part and multiply the fractional part by 2 again and you get the second bit to the right of the radix point as your integer part. Repeat this process until no fractional part remains (or until you have enough bits—many fractions go on forever). Figure 4.2 shows an example calculation.

1

The answers are 3.375 and 6.625.

Consider the decimal value 13

	Remainders	Quotients in binary:
2 | 13		1101.
2 | 6	1	110.
2 | 3	0	11.
2 | 1	1	1.
0	1	

Figure 4.1: Conversion of a decimal integer to binary.

Consider the decimal value 0.375

in decimal		in binary
0.375		0 .011
0.375 x 2 =	0 .75	0 .11
0.75 x 2 =	1 .5	1 .1
0.5 x 2 =	1	1

Figure 4.2: Conversion of a decimal fraction to binary.

Converting a fraction works on roughly the same basis as converting an integer. Each time you multiply by 2, you are effectively left shifting the equivalent binary fraction, making the bit to the right of the radix point into an integer, which is of course either zero or one. Because you are multiplying instead of dividing, the bits come off in the opposite order as when converting integers. It should be clear that you can use both procedures separately to convert a decimal number with both integer and fractional parts, and add the results.

Exercise 4.1.2 (decimal integer to binary). Convert the decimal value 22 to binary.[2]

Exercise 4.1.3 (decimal fraction to binary). Convert the decimal value 0.6875 to binary.[3]

Exercise 4.1.4 (decimal to binary). Convert the decimal value 14.1 to binary.[4]

---

[2] The answer is 10110.

[3] The answer is 0.1011.

[4] The answer is 1110.00011. Note that no binary number can represent the decimal value 0.1 exactly.

4.1.2   ADDITION

For addition, simply add bit by bit with carry. In the first (rightmost) column, we only need to be concerned with the binary sums $0 + 0 = 0, 0 + 1 = 1, 1 + 0 = 1$, and $1 + 1 = 10$. Only the last case can generate a carry into the next column (a "carry out"), leaving just the 0 in the first column. The second and later columns may potentially require the addition of three terms, because of the "carry in" from the previous column. This gives a new possible result, $1 + 1 + 1 = 11$, which gives a 1 in the current column and a carry out to the next. If the binary number is unsigned, and represented in a fixed number of bits, a carry out of the last column indicates the sum is too large to be represented in the available bits. As we will see later, carrying across the radix point makes no change in the procedure at all.

```
                          1   1                (carry)
         0   0   0   0    0   1   1   1        (augend)
    +    0   0   0   0    0   1   1   0        (addend)
         0   0   0   0    1   1   0   1        (sum)
```

Figure 4.3: The sum of two 8 bit binary numbers with carry digits shown.

4.1.3   SUBTRACTION

Subtraction can be performed directly using borrow. The borrow method works well with unsigned numbers as long as the result is positive. Computing columnwise from right to left, the cases $0 - 0 = 0$ and $1 - 0 = 1$ are fairly obvious, as is the case $1 - 1 = 0$, however the case $0 - 1$ seems to require a negative number. The solution is to simply "borrow" from the next higher column, adding two to the current column to give $10 - 1 = 1$. This operation leaves a 1 in the current column, and a "debt" of 1 in the next column. Since both the subtrahend and the borrow may need to be subtracted from the minuend, we now must also consider the situations of $0 - 10$ and $1 - 10$, which are of course handled by borrowing as well. As long as the minuend is larger than the subtrahend, the result will be positive. In unsigned arithmetic, a borrow out of the last column indicates a negative result and is an overflow since negative numbers are not representable. Signed arithmetic will be discussed shortly.

```
                     0   1    1   1   10        (borrow)
         0   1   1   0    0   0   0   0        (minuend)
    +    0   1   0   1    0   1   1   0        (subtrahend)
         0   0   0   0    1   0   1   0        (difference)
```

Figure 4.4: The subtraction of two 8 bit binary numbers with borrow digits shown.

4.1.4   MULTIPLICATION

When done by hand, binary multiplication is usually performed using a system of partial products. This algorithm is equivalent to breaking up the multiplicand into its power of 2 components, performing each single digit multiply as a shift, and then adding up the partial products. The algorithm works on the basis that the multiplicand b can be expanded as in (4.2):

$$ab = a(b_3 2^3 + b_2 2^2 + b_1 2^1 + b_0 2^0) = ab_3 2^3 + ab_2 2^2 + ab_1 2^1 + ab_0 2^0. \qquad (4.3)$$

In general, the product could require as many bits to represent as the sum of the number of bits in the multiplier and multiplicand. Consider the product of 0110 and 1011 shown in Fig. 4.5.

```
              0  1  1  0   (multiplier)
     ×        1  0  1  1   (multiplicand)
              0  1  1  0   (partial product when b₀ = 1)
           0  1  1  0      (partial product when b₁ = 1)
        0  0  0  0         (partial product when b₂ = 0)
     0  1  1  0            (partial product when b₃ = 1)
     1  0  0  0  0  1  0   (product)
```

Figure 4.5: The product of two four bit binary numbers with partial products shown.

Quite a few carries occur in computing the sum of the partial products, but these are omitted for clarity.

4.1.5   DIVISION

Unsigned integer division in binary proceeds just as it does in decimal. For example, The division of 150 by 13 results in a quotient of 11, and a remainder of $\frac{7}{13}$. Just as in decimal, the result could also be extended as a fraction beyond the radix point, but fractional arithmetic will be discussed later. In each of the subtractions shown below, several borrows occur, but they are not shown.

```
                       1  0  1  1   (quotient)
     1  1  0  1  )  1  0  0  1  0  1  1  0   (numerator)
                   1  1  0  1
                   0  0  1  0  1  1  1  0
                         1  1  0  1
                         0  1  0  1  0  0
                               1  1  0  1
                               0  0  1  1  1   (remainder)
```

Figure 4.6: The quotient of two binary numbers with remainder.

4.1.6   SIGNED BINARY REPRESENTATION

We have established a notation for representing unsigned integers and fractions in binary format. However, in signal processing, signed numbers are far more convenient. Three signed formats are usually discussed:

- Signed Magnitude,

- One's Complement, and

- Two's Complement.

Of these formats, two's complement is by far the most commonly used in hardware implementations.

Signed Magnitude   Signed magnitude representation is the format humans use on paper for most decimal calculations. It involves denoting the sign (humans use a $+$ or $-$ sign) separately from the magnitude. In computers, usually one bit is reserved for indicating the sign, with a 0 representing $+$ and a 1 representing $-$. Often, in a computer the most significant bit (MSB) is used for this purpose.

There are a few disadvantages to computing with this notation. A major disadvantage is that different algorithms are required for adding two numbers with like signs versus two numbers with opposite signs. Another problem is that there are two possible representations for zero ($+0$ and -0). This means testing for equality must be implemented carefully, and one of the possible represented values is wasted.

One's Complement   One's complement represents negative numbers as the binary complement of the positive number. In order to distinguish between positive and negative numbers, at least one bit (usually the MSB) must be reserved to always be zero for positive numbers. If the numbers are restricted to 8 bits, this would mean the MSB would effectively be a sign bit, the represented values would range from $2^7 = 128$ to $-2^7 = -128$, and there would again be both a $+0$ and a -0. Note the method of complements can be applied in decimal and has been used to implement subtraction since the days of mechanical calculators [15].

Two's Complement   The two's complement notation is widely used for computing, mainly because of its consistent treatment of addition of positive and negative numbers, but also because it has a unique representation of zero.

Computer hardware represents binary numbers using a finite number of bits, usually in "word" lengths of 8, 16, 32, or 64 bits, although almost any length is implementable. With a finite word length there will always be a limit to the range of values represented, and overflow behavior will be an important topic in subsequent sections. In unsigned integer arithmetic, if the wordlength is M, and the sum of two numbers is larger than can be represented, the result will just be the lowest M bits of the correct result.

For example, for 3 bit unsigned integers, the sum $3 + 5$ cannot be represented as the binary value for 8, because 8 is unrepresentable in this format. The truncated result would be zero.

$$
\begin{array}{ccccc}
 & 1 & 1 & 1 & \\
 & & 0 & 1 & 1 \\
+ & & 1 & 0 & 1 \\
\hline
 & 1 & \boxed{0 \quad 0 \quad 0} \\
\end{array}
$$

Figure 4.7: Binary addition resulting in overflow.

However, if we think of overflow as a useful result rather than an incorrect one, this behavior suggests that the two values just added can be thought of as negatives of each other, since their sum is effectively zero. If we choose to treat 011 as 3, then 101 has the property of behaving as -3. Fortunately, this behavior is quite consistent, so that adding $4 + (-3)$ produces a result of 1, and so on. A few moments of consideration will produce Table 4.1.

Table 4.1: A comparison of the three common signed binary formats for 3 bit integers.

Signed magnitude		1's complement		2's complement	
000	0	000	0	000	0
001	1	001	1	001	1
010	2	010	2	010	2
011	3	011	3	011	3
100	-0	100	-3	100	-4
101	-1	101	-2	101	-3
110	-2	110	-1	110	-2
111	-3	111	-0	111	-1

Number Circle   This system works because with a fixed number of digits (and ignoring overflow) the number system becomes circular. Repeatedly adding 1 simply moves counterclockwise around the circle, as seen in Figure 4.8. In two's complement, incrementing 111 and obtaining 000 is not an overflow, but a natural result. Now, however, incrementing 011 and obtaining 100 suggests $3 + 1 = -4$, which is clearly an overflow because 4 is not representable in this system.

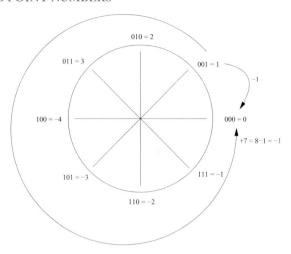

Figure 4.8: Number circle demonstrating subtraction in two's complement.

Addition and Overflow   Addition in two's complement can be done using hardware identical to that used in unsigned addition, because the operations are the same. However, because the meanings of the values are different, overflows are not the same.

Overflow in two's complement never occurs when adding two numbers of differing sign because the resulting magnitude gets smaller. When adding two numbers of like sign, the magnitude may become large enough to be unrepresentable, in which case the sign bit will change inappropriately. One simple way to detect overflow is to note cases when two like sign terms are added to produce a differently signed result.

It can be useful to detect overflow only on the basis of carry values. This can be done using only the carry into and out of the MSB column. Figure 4.9 shows all possible cases (a)-(h). In the case of two positive numbers, as in (a) and (b), the MSB column will be the sum of two zeros, and there will never be a carry out, but a carry in will cause the result to be negative (an overflow). In the case of two negative numbers, as in (c) and (d), there will always be a carry out, but a lack of carry in will cause the result to be positive (an overflow). This suggests that we should expect an *overflow whenever the carry in and carry out differ in the MSB column.* This will work because when the two inputs differ in sign, as in (e)-(h), the sum in the MSB column is $c + 0 + 1$, and the carry out will be equal to the carry in.

Exercise 4.1.5 (two's complement overflow).   Compute the following sums using 3 bit two's complement arithmetic, and check the overflow rules given above. (a) $3 + 2$, (b) $1 + 2$, (c) $3 + (-4)$, (d) $-2 + (-3)$, (e) $-2 + (-2)$.[5]

---

[5] (a) and (d) overflow.

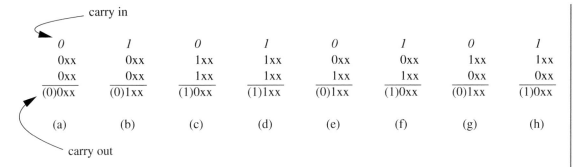

Figure 4.9: An illustration of overflow for all the possible cases in the most significant bit (MSB) when adding two's complement numbers.

Intermediate Overflow   When adding a list of numbers with mixed signs it is possible to have an overflow at an intermediate stage even if the final sum would be within range. As long as no special action is taken in this situation (such as saturating the result), the final sum will still be correct. This can be seen graphically by noticing that adding a positive number moves the sum clockwise around the number circle and adding a negative number effectively moves the sum counterclockwise around the number circle. As long as there is a net loop count of zero, the final answer will be correct.

Negation   There are two common methods for negating two's complement numbers: the complement and add one method, and the invert selected digits method. In Figure 4.10 you can see that inverting the bits of a binary number reverses the direction of increasing value and shifts the position of zero one position in the clockwise direction. Adding one shifts zero back one position in the counterclockwise direction, and the result is the exact set of negated values, except for the case of negating -4, which produces the incorrect result -4, an overflow due to the lack of representation for the value of $+4$.

The alternative (faster) method is to begin at the right most digit, and moving to the left, copy values until the first "1" is copied, then continue to the left, complementing the remaining digits. This method works because consecutive zeros in the right most bits will be complemented to ones, and the adding of a one will convert them back to zeros until a zero appears to stop the repeated carries. The zero will occur due to the first one in the digits moving right to left. The remaining digits are unaffected by the adding of one, and are thus left complemented.

Negation is used regularly in binary two's complement arithmetic, so it is worth the time to get familiar with the process of computing it.

Exercise 4.1.6 (two's complement negation). Use 8 bit two's complement to negate the values (a) 0000 0010 (b) 1101 1111 (c) 1100 0000.[6]

6
'0000 0010 = 64 = (64−)− (ɔ) 1000 0100 = 33 = (33−)− (q) 0111 1111 = 2− = (2)− (ɐ)

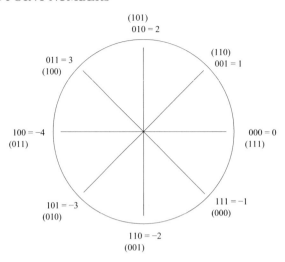

Figure 4.10: Number circle demonstrating negation in two's complement.

Sign Extension   Often we will need to modify the number of bits in a representation, e.g. convert an eight bit number to a sixteen bit number. In unsigned arithmetic, the new bits will always be zeros. With signed arithmetic, the MSB (the sign bit) is copied into the new bit positions. Consider the fact that -1 will always be all 1's regardless of the wordlength, so that adding one will "roll over" to zero. Likewise, if bits are removed, repeated leading ones or zeros can always be removed without fear of overflow, except for the last value, to preserve the sign.

Subtraction   Subtraction in two's complement can be accomplished either by negation and adding, or by using the borrow method. Overflow detection is similar to the methods used for addition — no overflow can occur when magnitudes are reduced, and the sign of the result should be obvious when the magnitude increases.

Multiplication with Negative Numbers   Multiplication in two's complement is identical to unsigned arithmetic if the two numbers are positive, but there are two ways to handle negative numbers. The first is to convert to positive numbers, and then correct the sign of the result (this is equivalent to using signed magnitude). The second method is to compute directly with negative numbers, however, because the result could require twice as many bits as the inputs, the sign extensions will not be correct unless the inputs are extended to the same bit length as the result before carrying out the arithmetic. Then the overflow behavior will be correct, but the processing will be rather inefficient. Many ingenious computational techniques exist to improve efficiency, but they are not covered here.

As an example of how the sign extension method works, consider the product $5 \times -3 = -15$. If we treat the two four bit two's complement inputs as unsigned numbers and compute the eight bit product, the result is 65, and if this is interpreted as a two's complement number, it would be -63, neither of which is the desired result.

It turns out that the lowest four bits of the result above are correct, and if we sign extend the four bit values to eight bit signed values, the eight lowest bits will be correct, including the sign.

```
            0  0  0  0  0  1  0  1   (multiplier)
  ×         1  1  1  1  1  1  0  1   (multiplicand)
                           1  0  1
                        0  0  0
                     1  0  1
                  1  0  1
               1  0  1
            1  0  1
         1  0  1
      1  0  1
      1  0  0  1  1  1  1  0  0  0  1   (product)
```

Figure 4.11: Signed multiplication using sign extension to the (double) length of the result. Only the lower eight bits of the product are kept.

If we discard all but the lowest eight bits, we obtain the desired result, a two's complement representation of -15.

4.2   Q-FORMAT

With only a few exceptions, arithmetic hardware treats all numeric inputs as integer values. However, the programmer is free to interpret the stored values as scaled constants, so that the stored integer really represents some combined integer and fractional value. The scale value is normally implied, not stored, and typically remains constant no matter how large or small the data values become. Since setting the scale value is equivalent to setting a radix point at some point other than to the right of the LSB, it is the constant position of the radix point (as compared to floating-point where each value has potentially different scale value stored as part of the number) which leads to the terminology "fixed–point" arithmetic.

Consider the values 0110 and 1001. As four bit two's complement integers, they represent 6, and -7, respectively. But, we can just as easily treat them as if they represent $bbbb \times 2^{-2}$, in which case they represent 1.5 and -1.75, instead. If we show the radix point explicitly, we might write 01.10 and 10.01, but in hardware, we would still compute as if the values were integers, and merely interpret the results based on the scale values, since the hardware is simplest using this method.

Since the implied scale factor can vary, engineers often use Q notation to specify the number of digits to the right of the radix point. Using Q-Format to describe the four bit value 01.10, we would call this a $Q2$ number. Note that traditional Q-Format does not specify the word length, but this is often implied by the wordlength of the processor architecture in a DSP μP environment. The most commonly used format for fixed-point arithmetic is $Q15$ format, with a 16 bit word. However, because of the flexibility afforded by FPGA architectures, and because our examples will often use much shorter wordlengths, we prefer to use a modified Q notation, specifying both the number of integer and fraction bits so that 01.10 would be described as $Q2.2$ format, or the typical $Q15$ format would be written as $Q1.15$. This notation has the advantage of being less ambiguous, and yet recognizable to practitioners used to the traditional format.

Still another format is in common use, e.g., in Matlab. Since it is possible to represent values where the radix point is entirely to the left or to the right of the available bits, either the number of integer or fraction bits can become negative when using the $Q\#.\#$ notation. For example, a 16 bit word would represent values in either a $Q-3.19$ or a $Q21.-5$ format, and as you can see, this notation becomes a bit unwieldy. Although numbers in these formats are rather uncommonly used, Matlab uses an abbreviated notation specifying the wordlength and the number of fraction bits, separated by a comma, and with a leading s or u to specify signed or unsigned format. The $Q1.15$ format would be specified as s16,15 in Matlab. The Matlab format also has the advantage of being unambiguous about the use of two's complement or not. Some examples of commonly used formats are shown in the Table 4.2.

Table 4.2: Three ways to express Q format parameters.				
Format Type	**Examples**			
Traditional Q format	$Q15$	$Q14$	$Q31$	$Q3$
Modified Q format	$Q1.15$	$Q2.14$	$Q1.31$	$Q1.3$
Wordlength, Fractionlength Format	s16,15	s16,14	s32,31	s4,3

To easily convert decimal values to Q format, simply multiply the decimal value by the scale factor 2^f where f is the number of fraction digits, round (or truncate), and convert the resulting integer to binary. If the value requires too many bits for the given format, you have an overflow, and if you get all zeros, you have an underflow. *Some texts use the term "underflow" to mean a negative value larger than the format will represent, while others use it to mean a value too small to distinguish from zero.* This text will use the term underflow to refer to values whose magnitude is too small to represent with anything but zero.

Example 4.2.1 (decimal to fractional). Convert the decimal value 0.3141592654 to a binary integer in the $Q1.15$ format.

To get an integer, we first scale by the required exponent and obtain

$$0.3141592654 \times 2^{15} = 10294.37080862720$$

then we can round the result and convert the integer 10294 to binary to obtain

$$0010100000110110.$$

Note that the result does not overflow since the value was less than one (technically $1 - \Delta/2$) and the left-most zero acts as a sign bit indicating a positive number. If we wish to view the binary result as a fractional quantity,

$$0.010100000110110$$

expresses the exponent 2^{-15} by placing the binary point 15 places to the left.

When a Q format number is written in binary with the radix point explicitly shown, the scaling is built into the position of the radix point, so no scaling is necessary, just conversion to or from binary. Of course, integer computer hardware will not recognize a radix point even if Q format numbers are occasionally written with one.

The procedure for conversion from Q format back to decimal is just the reverse: convert the binary integer to a decimal integer and then multiply by 2^{-f} to undo the scaling.

Example 4.2.2 (fractional to decimal). Convert the binary integer value 1011 to decimal assuming the format given is $Q1.3$.

First, recognize that the value given is negative due to the left-most bit being a one. Convert this to the corresponding positive integer of 0101. This binary value converts to a decimal integer 5. Undoing the scaling produces

$$-5 \times 2^{-3} = -\frac{5}{8} = -0.625.$$

Exercise 4.2.3 (decimal to fractional). Try converting the decimal values to binary integers in the format given (a) 1.3 to $Q2.14$, (b) 0.7 to $Q1.15$, (c) 0.8 to $Q1.3$ (d) -0.4 to $Q2.14$, and (e) -0.001 to $Q1.15$.[7]

Exercise 4.2.4 (fractional to decimal). Convert the binary integer values to decimal assuming the given Q format representation: (a) 1011 in $Q1.3$ (b) 0100101000111011 in $Q1.15$ (c) 0100101000111011 in $Q5.11$.[8]

4.3   FIXED–POINT ARITHMETIC

The primary purpose of Q format is to allow fractional data types on integer hardware. These fractional data types are supported in a few programming languages, including C extensions such as Embedded C. When fractional data types are used rather than floating point, a key concern of the programmer must be the range of the data values. If the data values are too large, overflow occurs, and if they are too small, the errors due to quantization can be large relative to the signal itself. It is also important to be aware of the changes in Q format required by the native operations of multiplication and addition.

7
8 (a) 0101001001100111 (b) 010110011001101010 0110 (c) 0110 (d) 11100110011001100110 (e) 111111011111111111

(a) -0.625 (b) 0.57992553710935 (c) 9.277808859375.

4.3.1   MULTIPLICATION

When two binary integers are multiplied, the result can be much larger than the operands. If we consider a multiplication operation $C = A \times B$, where the operands A and B have wordlengths of M_a and M_b, respectively, and fractional bit lengths of Q_a and Q_b, respectively, the necessary word lengths are given below. To avoid any potential overflow, the resulting wordlength should be $M_c = M_a + M_b$. The radix point will be correctly positioned if $Q_c = Q_a + Q_b$ as well.

Since wordlength cannot generally grow without limit, often it is necessary to truncate or round multiplication results to a shorter wordlength than what would prevent any loss of information. In fractional arithmetic, we prefer to lose the least significant bits since their loss causes the smallest errors. If we keep the most significant bits and reduce M_c, we get $Q_c = (Q_a + Q_b) - (M_a + M_b - M_c)$.

A common approach to fractional arithmetic is to operate entirely with magnitudes less than or equal to 1, so that datatypes such as $Q1.15$ and $Q1.31$ are all that are required for most operations. In this case, multiplies do not cause overflows (with the exception of the negation of -1), and reductions of wordlength should discard the rightmost, or least significant bits.

Though this approach is common, it is not without difficulties, as additions can still cause overflows and sometimes filter coefficients require representations of values larger than one. Also, it is still possible for the signal value to become too small and "underflow." When only a single integer bit is present, as in $Q1.15$, its value only determines the sign of the represented number, so it is essentially only a sign bit.

It is somewhat of a reversal of intuition to think in terms of keeping the signal near a magnitude of 1 (but not over) to preserve precision in the fraction bits when you compare this to the traditional integer approach taken in a typical C program, where lower bits are never lost, but care must be taken to avoid producing values too large for the datatype, and effectively, the most significant bits are always discarded. It is exactly this behavior in C that makes it difficult to express a multiply operation that starts with two low precision operands and produces a double precision result. The definition of C [14] requires the operands to be cast to the type of the result before the multiply occurs, which is quite inefficient in hardware.

Note that according to the rules described above, the product of two $Q1.15$ numbers will naturally produce a $Q2.30$ result, but we know that the result cannot be larger than 1, so a $Q1.31$ format would be sufficient. Since the conversion from $Q2.30$ to $Q1.31$ requires only a left shift of one bit, and the operation may be required often, some DSP μP's provide a mode or instruction allowing the left shift by one bit of every product. Such a mode is a form of processor specialization to enhance the efficiency of the architecture for fractional arithmetic.

4.3.2   ADDITION

When adding two fractional format numbers, the radix points must be aligned, and so usually the two operands are of identical format. When adding two numbers of the same format, the result can generally be twice as large as either operand, and therefore may require an extra integer bit to avoid overflow. Common solutions to the need for extra integer bits include the use of extra wordlength in the accumulator register where the sum is stored, or the scaling of the input values to assure that their sums remain less than the maximum representable value.

Using the notation above, we can define a general framework for wordlength in addition. If we consider the addition operation $C = A + B$, where the operands A and B have wordlengths of M_a and M_b, respectively, and fractional bit lengths of Q_a and Q_b, respectively, the necessary word lengths are given below. To align the radix points, we must require $Q_a = Q_b$ (this can be accomplished by zero padding if necessary). To avoid any potential overflow, the resulting wordlength should be $M_c = \max(M_a, M_b) + 1$. The radix point will be correctly positioned if $Q_c = Q_a = Q_b$.

Since wordlength cannot generally grow without limit, often it is necessary to truncate or round addition results to a shorter wordlength than what would be required to prevent any loss of information. For addition, this will often mean changing the radix point, or Q_c, to allow for larger results, and reduced precision. As with multiplication, we prefer to lose the least significant bits since their loss causes the smallest errors. If we keep the most significant bits and reduce M_c, we get $Q_c = Q_a - (M_c - \max(M_a, M_b) - 1)$, which is equivalent to computing
(New Q) = (intermediate Q)-(difference in word size).

4.3.3   ROUNDING

When reducing wordlengths, the simplest operation for most hardware implementations is to just discard the unnecessary bits. When discarding the least significant bits, this operation is often known as truncation. Truncation is desirable because it requires no extra hardware or operations. However, it does have the disadvantage of introducing a DC bias of $-\frac{1}{2}$ value of the least significant bit retained. This problem occurs because truncation always reduces the value of every input. Because both positive and negative values are shifted toward $-\infty$, the equivalent integer operation is called `floor()` in Matlab. The added DC bias in the resulting signal can cause problems with a variety of algorithms, notably the LMS adaptive filter to be discussed later. It can easily be removed with a simple high pass filter if necessary, although this may remove some of the computational advantage.

The main alternative to truncation is rounding. The most common form is usually achieved in hardware by adding half of the LSB to be kept before truncation. This results in values greater than half the LSB being rounded up by the carry out, and values less than half of the LSB being rounded down so that essentially no DC bias is introduced, but add and carry operations must be performed at every wordlength reduction. The equivalent integer operation in Matlab is `round()`.

4.4   AN FIR FILTER EXAMPLE

Suppose you want to implement a three point averager FIR filter using $Q1.3$ fractional arithmetic. Doing so will clearly illustrate many of the ideas discussed in the sections above about multiplication and addition in fixed-point.

The difference equation for a three point causal FIR filter can be expressed as

$$y[n] = \sum_{k=0}^{2} h[k]x[n-k] = h[0]x[n] + h[1]x[n-1] + h[2]x[n-2]. \tag{4.4}$$

To illustrate the effects of quantization clearly, we choose the three point averager, and input values which cannot be quantized exactly in $Q1.3$ format. The ideal values of $h[n]$, $x[n]$, and $y[n]$ are given below assuming implementation with no quantization. Note the limitation to four decimal digits is also a form of quantization.

$$
\begin{aligned}
h[n] &= [1/3 \quad 1/3 \quad 1/3] \\
x[n] &= [1/5 \quad 2/5 \quad 3/5 \quad 4/5] \\
\text{(ideal)} \quad y[n] &= [0.0667 \quad 0.2 \quad 0.4 \quad 0.6 \quad 0.4667 \quad 0.2667]
\end{aligned}
$$

Quantizing the values of $h[n]$ to a $Q1.3$ format can be done with rounding, so that the nearest available values are chosen. $Q1.3$ format contains values ranging from -1 to $+\frac{7}{8}$ in steps of $\frac{1}{8}$, so the value nearest to $\frac{1}{3}$ is $\frac{3}{8}$. The lines below show the unquantized value, the quantized value in decimal, and then the quantized values in binary so that the wordlength and radix point position for $Q1.3$ format are explicitly shown. Keep in mind that while the radix point is shown in these calculations, in most hardware implementations the value will be treated as an integer, and radix point interpretation is up to the algorithm designer.

		$h[0]$	$h[1]$	$h[2]$	
$h[n]$	= [	0.3333	0.3333	0.3333	]
$Q\{h[n]\}$	= [	0.375	0.375	0.375	]
$Q\{h[n]\}_{(2)}$	= [	0.011	0.011	0.011	]

Rounding the values of $x[n]$ into $Q1.3$ values requires rounding some up, and others down to reach the nearest representable value.

		$x[0]$	$x[1]$	$x[2]$	$x[3]$	
$x[n]$	= [	0.2	0.4	0.6	0.8	]
$Q\{x[n]\}$	= [	0.25	0.375	0.625	0.75	]
$Q\{x[n]\}_{(2)}$	= [	0.010	0.011	0.101	0.110	]

4.4.1   QUANTIZATION EXAMPLE - COMPUTING Y[0]

For the first output value, the convolution equation requires only the current input, since all past inputs are zero (initial rest assumption). Therefore, the computation required is simply

$$
y[0]_{(2)} \quad = \quad h[0]x[0] = \frac{\begin{array}{r} 0.011 \\ \times 0.010 \end{array}}{00.00\ 0\ 110} = 0.001 \quad \text{(rounded)}.
$$

$$y[0]_{(10)} \quad = \quad 0.125$$

In this computation you can see that the natural format of the product is $Q2.6$ as described in Section 4.3.1 the number of fractional bits is simply the sum of the numbers of fractional bits in the operands. To return the result to a $Q1.3$ format, only four bits can be retained. If we were truncating results, $y[0]$ would be zero, but since we are rounding in this example, the result is rounded up. Recall that the ideal value is $y[0] = 0.06667$, and note that the nearest $Q1.3$ value to the ideal result is 0.125.

4.4.2   QUANTIZATION EXAMPLE - COMPUTING Y[1]

$$y[1]_{(2)} = \quad h[0]x[1] \quad + \quad h[1]x[0]$$

$$
\begin{array}{cc}
0.011 & 0.011 \\
\times 0.011 & \times 0.010 \\
\hline
00.00\ \mathbf{1}\ 001 & 00.00\ \mathbf{0}\ 110
\end{array}
$$

$$
\begin{array}{r}
00.001001 \\
+00.000110 \\
\hline
00.\mathbf{011}111
\end{array}
$$

$$y[1]_{(2)} = 0.010 \quad \text{(rounded)}$$
$$y[1]_{(10)} = 0.25$$

In the computation of $y[1]$, you can see that both multiplies produce double precision (8 bit) results with the same Q format. If the addition is performed at the full product precision, the effects of the lower bits have an opportunity to accumulate and effect the upper bits that will appear in the four bit result. This is a more accurate result than if each product were rounded to $Q1.3$ before the addition. A statistical model of this effect will be discussed later. Accumulation with extended precision is common in DSP microprocessors.

4.4.3   QUANTIZATION EXAMPLE - RESULTS

If any doubt remains about how these computations are implemented, the reader should verify the results for $y[2]$ through $y[5]$ by implementing (4.4) by hand as shown above for $y[0]$ and $y[1]$. The correct results are shown below.

computed	$y[n] = [$	0.125	0.25	0.5	0.625	0.5	0.25	$]$

ideal	$y[n] = [$	0.0667	0.2	0.4	0.6	0.4667	0.2667	$]$

quantized ideal	$y[n] = [$	0.125	0.25	0.375	0.625	0.5	0.25	$]$

It may be a bit surprising that the computed result and the quantized version of the ideal result differ only by one quantization step for one sample time. Hopefully, this example will serve to build confidence that the fixed-point arithmetic shown can be quite useful even with limited precision.

Of course, we will be exploring the limitations of fixed-point arithmetic in later chapters. You may also have noticed that no overflow was encountered in this example despite the lack of extra integer bits, mainly because of the size of the input signal and filter coefficients. A systematic approach to the design choices influencing overflow will also be discussed later.

4.4.4   MATLAB EXAMPLE

Although the four-bit computation example given above is instructive to perform by hand, it is also useful to perform such a simulation in Matlab, to introduce the high level tools available for setting the arithmetic parameters so that the exact results can be obtained with far less effort.

The example above can be implemented using the `dfilt` object in Matlab. To use the `dfilt` objects, the Filter Design Toolbox is required, and to use them to simulate fixed-point arithmetic, the Fixed-Point Toolbox is required.

In the code example given below, the routine `setq3()` is the only function used that is not a part of the toolboxes, and its function is to set the formats used for signals, coefficients, products, and sums in the `dfilt` object.

```
%% fourbitfilter.m
% This file is a demo of the four bit by-hand example computed in Matlab
% using a discrete time filter object.


%% set up the fixed-point filter
b = [1/3 1/3 1/3]; % start with a 3 point averager FIR filter
Hd = dfilt.dffir(b); % create a dfilt object, which defaults to 'double'
                     % arithmetic
Hd.arithmetic = 'fixed'; % change the dfilt to fixed point arithmetic,
                         % which defaults to 16 bit
                         % word lengths, and automatic fraction lengths
specifyall(Hd); % eliminate all automatic scaling of values,
                % since we want to study the choices not ignore them

Hd = setq3(Hd,[4 3 8 7]);  % my helper function to set data and
                           % coefficients to s4,3 and accumulators to
                           % s8,7


%% look at the filter properties
% The filter properties associated with fixed-point don't show up
% until you convert the arithmetic type to fixed.

Hd
```

```
%% run the filter
% The input will be quantized, then the output will be computed using
% quantized coefficients (wordlengths set above).
x = [1/5 2/5 3/5 4/5 0 0];
y = filter(Hd,x) % the output should be the same as we saw working by
                 % hand
```

4.5   FLOATING–POINT

A complete discussion of floating point precision effects is outside the scope of this text. However, a brief description of the format is appropriate, and makes it easier to make occasional comparisons between fixed–point and floating–point.

The most commonly used floating point formats are described by the IEEE 754 standard [8, 6], and are commonly used by the C language data types, *float*, and *double*. In a 32 bit architecture, the float, or single precision floating–point value uses 23 bits for the fractional part, 1 bit for the sign, and 8 bits for the signed exponent value. The 8 bit exponent is signed using a shift-by-127 representation that is not two's complement. The exponent bits allow exponents from 2^{-126} to 2^{127} with some values lost to exceptions for zero, *Inf*, and *Nan*, etc. The fractional part is usually normalized to place the radix point to the right of the most significant 1 bit (magnitudes are always positive), so by assuming the 1 bit to the left of the radix point, there are effectively 24 bits of precision available.

To compare IEEE floating point to the fixed-point Q notation we have established, consider the approximation of single precision floating point as a 24 bit fixed-point number with a variable Q point that runs from -126 to 127. Because of the various exceptions required by the standard, this is only an approximation, but an adequate one for our purposes of comparative precision. Table 4.3 shows how single and double precision floating–point formats compare in bit allocation and precision.

Table 4.3: Comparing single precision and double precision floating point formats. [2]

Property	Single Precision Float	Double Precision Float
Wordlength	32	64
Effective Fraction Bits	24	53
Exponent Bits	8	11
Max Exponent	127	1023
Min Exponent	-126	-1022
Max Value	$(2 - 2^{-23}) \times 2^{127}$	$(2 - 2^{-52}) \times 2^{1023}$
	$= 3.403e{+}038$	$= 1.798e{+}308$
Min Positive Value	$2^{-23} \times 2^{-126} = 1.401e{-}045$	$2^{-52} \times 2^{-1022} = 4.941e{-}324$
Epsilon	$2^{-23} = 1.192e{-}007$	$2^{-52} = 2.220e{-}016$

The most important distinguishing property of floating point arithmetic is that the exponent (or Q point) of each value can be adjusted to allow the maximum precision, and the necessary exponent is stored with the fractional data. This system allows for good precision for both large and small numbers, but there are disadvantages, too. One disadvantage is that hardware to support the separation and recombination of the fractional and exponent parts requires extra silicon area, gate delay time, and power, or if these functions are performed in software, they slow down processing substantially.

Although Moore's Law [21] is diminishing speed and silicon area differences, wordlength alone can also be important to system cost because data memory sizes have a large effect on power and cost. You may have noticed that there is no common 16 bit floating point data type, but the fixed–point 16 bit $Q15$ format is quite common. This is because the floating point type must divide the available bits between two numbers (fraction and exponent), limiting the range of both values for short wordlengths. Since a short wordlength can help reduce system cost, fixed–point has remained popular despite the greater ease of floating–point development.

Another method of improving accuracy over wide dynamic ranges for short wordlength is the use of log-spaced quantization values. Although this method is used in the telephone system [9] for μ-law and A-law encodings to improve the performance of 8 bit word length data, it is relatively expensive to process this data, since adding and subtracting are no longer straightforward.

4.6   BLOCK FLOATING–POINT

In most hardware architectures there is only support for integer arithmetic. In systems where the dynamic range of fixed-point integer arithmetic is insufficient, it may be justifiable to include floating-point arithmetic support. Since floating–point hardware has significant cost and speed disadvantages, one may wonder if there is a compromise available.

Block floating–point is a powerful technique that preserves the wordlength and computational efficiency of fixed–point, while allowing many of the advantages of floating–point. Since many signal processing tasks are computed on blocks of data to minimize loop overhead, significant improvements in computational accuracy can be obtained by determining the maximum value in a block of data, and then scaling all the values up (usually by a left shift) to the maximum range without overflow. The shift represents a power of 2, and also a scaling exponent, just as in floating–point, but there is only a single exponent for the entire block. Therefore, the values in the block can still be used in computations as fixed–point values and many more bits of precision are available for small values when they are scaled up to occupy the most significant bits of the data word.

When an operation requires a larger range, the values can be shifted down, and the exponent adjusted accordingly. The method adapts with the signal, like floating–point, but only one block at a time. Block floating–point is most effective for handling signals that have a wide range of signal levels, but the levels tend to be similar over a single block.

The disadvantages of block floating–point are that every sample in the block must be tested for maximum values whenever the exponent needs to be set, and the advantages are minimal if it is important to accurately handle both large and small value ranges within the same block.

Quantization Effects: Data and Coefficients

5.1   FOUR TYPES OF ERROR

In a fixed–point system, the signal values are always an approximation of the values that would appear in an ideal–precision system (remember that floating–point systems are not truly ideal, though they are normally closer than a fixed–point system). If we handled all the differences between an ideal system and a fixed–point system as one combined error source, analysis would be overwhelming. Instead, we consider error sources separately according to how they can be modeled and analyzed. This approach makes it practical to predict and minimize each of the non-ideal effects considered.

Four (4) kinds of error are introduced in a fixed point system:

1. Input quantization (model as noise — quantization noise)

2. Coefficient quantization (easy to simulate)

3. Product quantization (round-off error, also called underflow—modeled as noise)

4. Overflow (trade-off between overflow & round-off error)

For each type of error, we will consider how errors are introduced, how they can be modeled, and how to minimize the effects on the system's overall performance. One section of this chapter is devoted to each of the error types listed above. Although input quantization and product quantization are both modeled as noise sources, they each introduce error for slightly different reasons, and they require different kinds of analysis, so they are discussed separately.

Although the quantization process is deterministic and nonlinear, linear analysis can be used if the quantization errors introduced are modeled as additive noise. Since it is often useful to model the signal as a random process, this analysis is both practical and commonplace. Coefficient quantization is usually considered as a fixed error in the frequency response of the filters to be implemented. Since the coefficients are usually known in advance, the actual frequency response can be tested and adjusted if necessary.

Product quantization happens because whenever the signal is multiplied by a non-integer constant value, additional fractional bits are necessary to accurately represent the result, and so the ideal product will likely be quantized again to prevent unlimited growth in the wordlength. Fortunately, product quantization can be treated similarly to input quantization, the only difference is the need to account for noise injected inside the filter flowgraphs.

Overflow problems occur because fixed point values only represent a finite range, and the signal of interest may grow beyond that range. Preventing overflow is possible by scaling the input signal down, but this causes a loss of precision. In terms of the quantization noise model, there is a tradeoff between the likelihood of overflow when the signal is large, and the relative power of the quantization noise if the signal is small. This tradeoff is built-in to the fixed–point implementation because both range and precision are limited in the fixed–point representation.

5.2   DATA QUANTIZATION

Real world signals such as audio from an analog microphone, are continuous both in time and in value. Processing them with a computer requires that they be converted into a representation that is discrete, both in time (sampling) and in value (quantization). The hardware device that performs this function in most systems is an analog to digital (A/D) converter. The internal functions of A/D converters involve signal processing, digital, and analog electronics, and their specifics are beyond the scope of this text. We will focus instead on the effects of converting a signal into discrete values.

5.2.1   ANALOG-TO-DIGITAL CONVERSION

The purpose of a quantizer is to convert an infinite number of input values into a finite number of symbols, usually represented by a binary word. We have already discussed the unsigned and two's complement orderings for a binary word, but both provide the same number of symbols. The most common arrangement is for output symbols to represent any of the input values within a range of $[x - \Delta/2, x + \Delta/2]$ around the symbol's "ideal" value, x. This provides a linear symbol spacing, and a constant Δ.[1] Since a finite number of symbols is available and each symbol represents a finite range, the useful range of the quantizer is also finite, and values outside its representable range will have a large error and are said to be in saturation.

An A/D converter that converts an equal range of values above and below the ideal value of the resulting symbol is a **rounding** quantizer, but there are cases where a **truncating** quantizer can be useful. A truncating quantizer converts a range around the ideal value of $[x + 0, x + \Delta]$ to the symbol representing x.

Assuming there are B bits in the quantized symbol, the range can be divided into 2^B levels. Often these levels are evenly divided between the positive and negative input values, but some implementations use only positive values, which can simplify the hardware design. Figure 5.1 shows the input to output characteristic for a simple rounding quantizer.

A good example of the quantization problem is the behavior of a CD player. A continuous audio signal must be quantized to 16 bit values for storage on a CD. Since audio line-out levels are generally limited to 1 V_{p-p}, a good quantizer choice would be to divide the range $[-\frac{1}{2}V, +\frac{1}{2}V]$

---

[1]It is quite difficult to implement a high resolution A/D converter that performs nearly ideally. Real A/D converters are rarely perfectly linear, and at high resolutions it is common for some symbols to be skipped because the least significant bits are difficult to keep precise across a wide range of values. In some designs, the least significant bits are meaningless due to noise. These non-ideal behaviors are important, but outside the scope of this text.

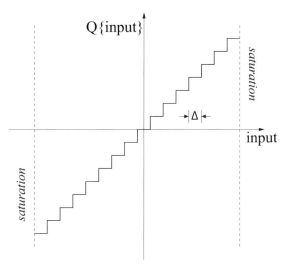

Figure 5.1: Output vs. input for a linearly spaced rounding quantizer.

into 2^{16} ranges of width $\Delta = \frac{1V}{2^{16}}$. If the range containing zero is centered around zero (rounding), there is an odd number of levels left, and the remaining levels cannot be perfectly symmetric around zero.

When a music signal is quantized, each quantized value can be completely determined from the input to the quantizer, such that $x[n]$ becomes $Q\{x[n]\}$. The quantization operation $Q\{\}$ is memoryless, deterministic, and as is obvious from Figure 5.1, it is nonlinear. Since non-linear analysis is extremely difficult, it is far more practical to handle quantization by modeling the quantization process as a linear process.

Instead of a non-linear process, we can treat the quantization as the addition of an error signal, and approximate the error as a uniform white noise signal. As anyone who has listened to the difference between a quantized and an unquantized audio signal would know,[2] quantization error sounds like white noise. The addition of white noise can be handled with linear analysis and so all the traditional systems analysis tools are at our disposal.

$$Q\{x[n]\} = x[n] + e[n] \tag{5.1}$$

For a rounding quantizer, the error values $e[n]$ fall in the range of $[-\Delta/2, \Delta/2]$, and for a truncating quantizer, the error values fall in the range $[-\Delta, 0]$.

---

[2]Take a moment to do this if you have access to the text website at www.morganclaypool.com/page/padgett or http://www.rose-hulman.edu/~padgett/fpsp. Use the Matlab file quantization_error_audio.m to listen to quantization noise.

Quantization Model   We model quantization (non-linear) as the addition of noise (linear). To allow for a simple model, we make the following assumptions:

- $e[n]$ is uniformly distributed on $[-\Delta/2, \Delta/2]$ (rounding)

- $e[n]$ is white noise

- $e[n]$ is uncorrelated with $x[n]$

- $e[n]$ is a zero-mean (for rounding), stationary process

Modeling $e[n]$ as a uniformly distributed, white random process only makes sense if values of quantization error are unrelated to each other, and usually equally likely. Quantization error values do have these properties in many situations, but not all, and therefore we will describe some constraints that make the model more reliable.

The model is only (somewhat) valid if we require:

- Quantization steps must be sufficiently small relative to the signal amplitude.

- The signal must vary sufficiently.

    – The signal should traverse several quantization levels between two successive samples.

- There must be no saturation or overflow.

- The signal $x[n]$ must not be periodic with a period that is an integer multiple of the sample rate.

If quantization steps are large compared to the signal amplitude, the error signal begins to look like small segments of the signal itself as seen in Figure 5.2, and less like a random process. Even if the signal amplitude is large, it may vary so slowly that successive values all quantize to the same level, and then the error again looks more like the signal than like a random process. However if the signal varies rapidly over several levels as in Figure 5.3, the error value has little relationship to nearby values, and any value of error is just as likely as any other. This produces an error signal very similar to a white, uniform distributed random process. Widrow and Kollár [28] go into a great deal of detail on the validity of the quantization noise model.

Saturation and overflow effects cause errors larger than $\Delta/2$, and therefore cause the distribution to be nonuniform. If $x[n]$ is periodic with a period that is an integer multiple of the sample rate, then the signal values will repeat and so will the quantization values of $e[n]$. Then, instead of being a spectrally white signal, $e[n]$ becomes periodic and all its power is in harmonic terms, so that it fits the given noise model very poorly indeed. In the best case, the statistical model is only an approximation, but for many systems it predicts observed behavior quite well, and often works well even when the given constraints are not fully met.

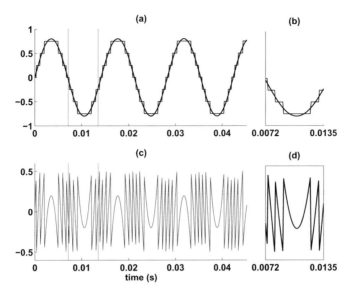

Figure 5.2: A sinusoid quantized with so few levels that the quantization error looks like the signal. Parts (a) and (b) show the signal and its quantized version, while parts (c) and (d) show the quantization error normalized to a quantization step.

5.2.2   RANGES

The full-scale range, R_{FS} for a 2's complement representation is given by:

$$- (2^B + 1)\frac{\Delta}{2} < x_{input}(nT) \le (2^B - 1)\frac{\Delta}{2} \tag{5.2}$$

where

$$\Delta = \frac{R_{FS}}{2^B} \tag{5.3}$$

Equation (5.2) is easily misunderstood if you only think of the ideal quantization levels themselves. The minimum and maximum quantization levels are

$$\frac{-2^B \Delta}{2} \text{ and } (\frac{2^B}{2} - 1)\Delta,$$

respectively. These levels do not represent the full range of quantizer because values $\frac{\Delta}{2}$ above the maximum and below the minimum ideal values are just as valid as those between the values.

The quantization error $e[n]$, is bounded in amplitude according to

$$-\frac{\Delta}{2} < e[n] \le \frac{\Delta}{2}. \tag{5.4}$$

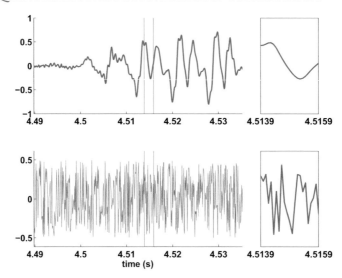

Figure 5.3: A speech signal (sampled at 11025 Hz) quantized with 12-bits. The quantization error is shown below (zoomed at right) normalized to a quantization step.

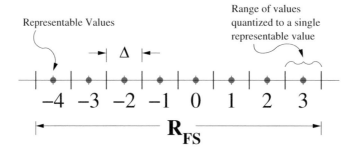

Figure 5.4: A quantization scheme with $B = 3$ and $\Delta = 1$. The $R_{FS} = 8$, and the ideal quantized values (small marks) are integers. Note that the boundaries for each quantized level (large marks) occur halfway in between integers.

A plot of the $e[n]$ of a 3-bit A/D converter as a function of the input sample $x[n]$ is shown in Figure 5.5. Because two's complement values are not symmetric around zero, neither is the full scale range.

Assuming a data range of $(\pm 1) - \frac{\Delta}{2}$ (as in 16-bit $Q15$), the step size is

$$\Delta = \frac{2}{2^{16}} = 2^{-15}.$$

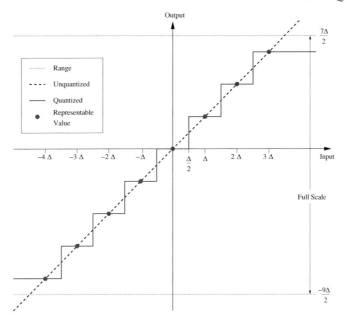

Figure 5.5: The values up to $\frac{\Delta}{2}$ above and below the ideal quantized levels produce quantization errors within the rounding range.

Rounding to a quantized value introduces errors in the range

$$\pm\frac{\Delta}{2} = \pm 2^{-16}.$$

5.2.3   QUANTIZATION NOISE POWER

To calculate the average power in a random process $e[n]$, one can compute the expected value of the square, or $E\{e^2[n]\}$, which is also the second moment. Since $e[n]$ assumed to be stationary, the definition of the second moment is

$$E\{e^2[n]\} = \int_{-\infty}^{\infty} x^2 p_e(x)dx. \tag{5.5}$$

Recall that x is simply a dummy variable to allow integration over the range of the PDF (probability density function), $p_e(x)$. Since $e[n]$ is zero mean the second moment is the same as the variance. Since the noise is assumed to be white and uniform on $[-\Delta/2, \Delta/2]$, $p_e(x)$ has a constant value of $\frac{1}{\Delta}$ over that interval, and is zero otherwise.

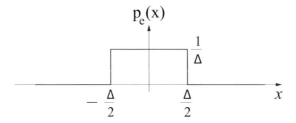

Figure 5.6: The probability density function of zero mean uniformly distributed error.

So, the average power can be calculated as

$$\sigma_e^2 = \int_{-\Delta/2}^{\Delta/2} x^2 \frac{1}{\Delta} dx$$
$$= \frac{\Delta^2}{12} \tag{5.6}$$

And substituting in our expression for Δ we have:

$$\sigma_e^2 = \frac{\left(\frac{R_{FS}}{2^B}\right)^2}{12} = \frac{R_{FS}^2}{12 \cdot 2^{2B}} \tag{5.7}$$

$$\sigma_e^2 = \frac{\Delta^2}{12} = \frac{2^2}{2^{32} \cdot 12} = \frac{2^{-30}}{12} \quad \text{for } Q15$$
$$= 77.6 \times 10^{-12} \quad \text{(16-bit } Q15 \text{ noise power)}$$
$$= -101.1 \text{ dB}$$

The 16-bit $Q15$ quantization noise power is **fixed** as long as the basic assumptions about the signal behavior are met. Remember that although there is no quantization noise without a signal present, there is a minimum quantization noise power associated with typical signals.

5.2.4   SIGNAL-TO-NOISE RATIO

The ratio of the signal power to noise power, or signal-to-noise ratio (SNR) is a common and important measure of system performance.

$$\text{SNR} = \frac{\text{signal power}}{\text{noise power}}$$

This is usually expressed in decibels as

$$\text{SNR} = 10 \log_{10} \frac{\text{signal power}}{\text{noise power}} \quad \text{(dB)} \tag{5.8}$$

We know how to estimate the noise error based on the quantizer characteristics. The other part of the SNR is based on signal power.

Example 5.2.1 (SNR). Consider a sinusoidal input:

$$x[n] = A \cos(\omega n)$$

If we want to find the SNR due to quantization noise in $Q\{x[n]\}$, we simply need to compute the signal power and divide it by the quantization noise power. This is consistent with our model of quantization noise as additive white noise, or $Q\{x[n]\} = x[n] + e[n]$. Average power for a sinusoid is

$$\text{signal power} = \frac{A^2}{2}.$$

Now we can find the SNR for a full-scale sinusoidal input

$$
\begin{aligned}
\text{Full Scale Sinusoid SNR} &= \frac{A^2/2}{\Delta^2/12} \\
&= 6\frac{A^2}{\Delta^2}
\end{aligned}
\tag{5.9}
$$

For a full-scale sinusoid, in $Q15$ format,

$$\text{SNR} = \frac{6}{2^{-30}} = 6.44 \times 10^9 = 98.09 \text{ dB SNR}$$

This is roughly the SNR obtained by a CD player because its SNR is limited by the same quantization noise power. However, music signals have a wider bandwidth and lower average power than a full scale sinusoid, so CD players usually specify a lower SNR. [3]

SNR per Bit for a Full-Scale Sinusoid We can rearrange the relationship above to emphasize the role of the number of bits used to express a sinusoid. Given B bits, and $R_{FS} = 2$, we have $\Delta = \frac{2}{2^B}$, so

$$
\begin{aligned}
\text{Full-Scale Sinusoid SNR} &= 6\frac{A^2}{4/2^{2B}} \\
&= \frac{3}{2}A^2 2^{2B}.
\end{aligned}
\tag{5.10}
$$

$$
\begin{aligned}
\text{Full-Scale SNR (in dB)} &= 10\log_{10}(\frac{3}{2}A^2) + 2B\log_{10}2 \text{ dB} \\
&= 20\log_{10}(A) + 6.021B + 7.7815 \text{ dB}
\end{aligned}
\tag{5.11}
$$

Obviously, A must not cause overflow, so for this example we require $|A| < 1 - \frac{\Delta}{2}$. Equation (5.11) makes it clear that increasing the number of bits by one will improve the SNR by 6 dB, and decreasing the signal power will reduce the SNR directly.

---

[3] SNR for the entire band can be misleading if the signal is narrowband and the noise power is spread over the entire band. In this case, SNR in just the band of interest is much more useful.

SNR per Bit for General Case   For signals modeled as a random process, or where only the signal power σ_x^2 is known, the SNR can be expressed in terms of the number of bits, and the ratio of the standard deviation σ_x to the full scale range R_{FS}. For zero mean deterministic signals, the standard deviation is analogous to the RMS value of the signal.

$$SNR = 10\log_{10}\left(P_x/P_e\right)$$

where for random processes, $P_x = \sigma_x^2$ is the signal power and $P_e = \sigma_e^2$ is the noise power. Substituting (5.7) we have

$$
\begin{aligned}
SNR &= 10\log_{10}\left(\frac{\sigma_x^2}{\frac{R_{FS}^2}{12\cdot2^{2B}}}\right) \\
SNR &= 10.792 + 6.021B + 20\log_{10}\left(\frac{\sigma_x}{R_{FS}}\right)
\end{aligned}
\tag{5.12}
$$

The ratio of the peak value (roughly $\frac{R_{FS}}{2}$) to the RMS signal value is an important quantity known as the peak to RMS ratio (PRR), and it is sometimes expressed in dB as *crest factor*. The implications of signal crest factor on system behavior will be discussed further in the scaling section below. Equation (5.12) also makes it clear that SNR increases by about 6 dB for each additional bit in the signal representation.

5.2.5   SATURATION AND OVERFLOW

So far we have ignored a large problem with quantization: **overflow**. Overflow occurs when the signal to be quantized exceeds the maximum positive or negative quantization levels. Most quantization systems do not have a way to signal that the input is out of range, and so they must produce an erroneous output value.

In two's complement, there are two common ways to respond to out of range input values: **saturation**, and **wrapping**. In saturation, the out of range value is replaced with the nearest valid value, either the maximum positive value or the maximum negative value. In wrapping, the value returned is just the least significant bits of the correct value (assuming the value could be represented correctly using a larger wordlength).[4]

Saturation produces better signal behavior than wrapping in an overflow situation because it minimizes the size of the error. When wrapping, a signal a single step above the maximum positive value will produce an output of the maximum negative value — a tremendously larger error value. However, when quantization occurs in arithmetic operations, the hardware necessary for saturation behavior is significant, and wrapping occurs naturally, so both methods may be useful depending on system requirements.

[4]Because arithmetic overflow in an add operation produces wrapping behavior, wrapping is sometimes referred to as just "overflow" and considered distinct from saturation, but the terms "wrapping" and "saturation" are more specific.

Saturation in A-to-D Converters   Most A-to-D converters operate on input signals which have unpredictable peaks. For example, if a signal has a Gaussian noise component, the peaks can (with low probability) reach any finite value. If we wish to avoid overflow events for such a signal, we must scale the input signal down to reduce the likelihood of peaks large enough to overflow. However, since the quantization noise power is fixed (for a given quantizer), the signal power cannot be reduced without reducing the SNR.

Therefore, the input to a quantizer (for example, an analog-to-digital converter) must be scaled so that

- The signal power of the quantized signal is maximized while

- The probability of saturation is minimized.

These are conflicting constraints, and so some analysis of the tradeoff is necessary.

For intuition, we observe that in professional audio, a rule of thumb is to keep the maximum VU Meter spike at 6dB below full-scale. This is simply one possible choice of compromise between saturation and SNR, for a specific type of signal.

Probability of Overrange Errors   Because so many natural signals have Gaussian behavior (often due to the Central Limit Theorem [24]) and because the analysis is straightforward, we often model the input signal as a Gaussian random process. In this case we can calculate the probability of having an overrange error by simply integrating the area under the PDF outside the valid range.

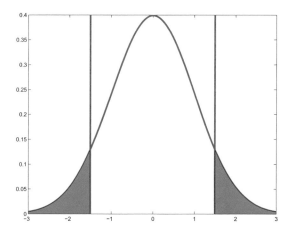

Figure 5.7: A Gaussian PDF with the region used to compute the probability of overflow shaded.

If the input signal $x[n]$ is assumed to have Gaussian distributed samples we can calculate the probability of having an overrange error. We assume a full-range of $K\sigma_x$, which implies a PRR of $K/2$ since the peak values are $\pm\frac{K\sigma}{2}$, and the RMS value is σ.

To calculate the probability of overflow for a Gaussian, we use the standard normal cumulative distribution function, defined as

$$\Phi(\alpha) = \frac{1}{\sqrt{2\pi}} \int_{-\infty}^{\alpha} e^{-y^2/2} dy \tag{5.13}$$

which can be thought of as the probability that a zero mean, unit variance Gaussian random variable will be less than α. Since the quantizer can overflow whenever $|x[n]| > \frac{K\sigma}{2}$, we need

$$P(\text{overflow}) = 2(1 - \Phi(K/2)) = \frac{2}{\sqrt{2\pi}} \int_{K/2}^{\infty} e^{-y^2/2} dy. \tag{5.14}$$

Note that $1 - \Phi(K/2)$ is simply the probability that $x[n] > \frac{K\sigma}{2}$ normalized to the unit variance case, and by symmetry, the probability of overflow is doubled. The results of (5.14) can be obtained in Matlab using the expression P = erfc(K/(2*sqrt(2))).

Table 5.1: Probabilities of overflow for various values of K.

K	4	5	6	7
P(overflow)	0.04550	0.01242	0.00270	0.00047
K	8	9	10	
P(overflow)	0.6334e-4	0.6795e-5	5.7330e-7	

SNR vs. Probability of Overrange   As noted above, reducing the probability of overflow also reduces the SNR. For various values of K and B, we see that SNR increases with increasing B, and decreases with increasing K [20]. The SNR is given in dB by the expression

$$\text{SNR} = 10 \log_{10} \frac{12 \cdot 2^{2B}}{K^2}. \tag{5.15}$$

Table 5.2: SNR in dB for a Gaussian signal with B bits and specified K. [20]

	$B = 8$	$B = 10$	$B = 12$	$B = 16$	$B = 32$
$K = 3$	49.41	61.46	73.50	97.58	193.91
$K = 5$	44.98	57.02	69.06	93.14	189.47
$K = 7$	42.05	54.10	66.14	90.22	186.55
$K = 9$	39.87	51.91	63.95	88.03	184.37

These two tables can be used to estimate bit requirements for signal representation. These estimates will be exact for Gaussian signals, but only approximate for other signal types.

Crest Factor   Although Gaussian signal modeling is useful and powerful, many signals have value distributions that are not remotely Gaussian. Speech signals for example, usually have large peaks compared to their average power. In terms of the distribution function, this would be considered a "heavy-tailed" distribution. Narrowband or sinusoidal signals, on the other hand, have relatively high signal power compared to their peak values. Since peak values determine clipping behavior, and signal power effects SNR, we will characterize signals in terms of their peak to RMS value ratio (PRR), or equivalently, their crest factor (CF). The crest factor is defined here as

$$CF = 10\log_{10}\frac{pk^2}{S} = 10\log_{10}\frac{(2^{B-1}\Delta)^2}{S} \tag{5.16}$$

where pk is the signal peak absolute value, S is the signal power, B is the number of bits, and Δ is the quantization step size. Therefore, CF is just the ratio of pk^2 over signal power in dB. Using $20\log_{10}(PRR)$ would give the same result.

When varying the CF, what matters is the ratio of pk^2 to S, so we can choose to fix all the parameters but one, and use the one to adjust the CF. Because each of the several obvious choices gives a different way of viewing the tradeoffs of scaling, etc., we will examine them here.

- If we adjust S to obtain various CF values, and hold B, Δ, and therefore pk constant, the effect is similar to scaling the signal up and down while keeping the same signal format, such as $Q1.15$. Since Δ is constant, quantization noise power is constant, and S controls the SNR, i.e. increasing CF reduces S, reducing SNR.

- If we adjust Δ while holding B and S constant, pk and Δ change together. The effect is similar to adjusting the exponent in a format such as QM.(B-M). Here the quantization noise increases and SNR decreases with increasing CF.

- If we adjust B while holding Δ and S constant, pk increases with B. This result could be obtained by, for example, treating the data format as all integers, or QB.0. Now that S and Δ are constant, the SNR is constant. Note that although S is constant, we can set it to any value we like, and thus obtain any desired SNR.

A CF of 0 dB without clipping is only possible when the signal power is equal to pk^2, and therefore only a squarewave-like signal (or DC) could have a CF of 0 dB without clipping.

Now we define

$$clip_{pk}\{x[n]\} = \begin{cases} x[n] & \text{if } |x[n]| < pk \\ pk & \text{if } |x[n]| \geq pk \end{cases} \tag{5.17}$$

which is just a saturation operator. We may now define the clipping error as

$$e_{clip}[n] = clip_{pk}\{x[n]\} - x[n]. \tag{5.18}$$

If we take several example signals with different distributions and compute the power in e_{clip} at different CF values we obtain Figure 5.8 [26]. Note that the S for each signal need not be the same, since CF is determined by the ratio, and the clipping noise power is in dB relative to S.

Figure 5.8 shows that clipping error power depends heavily on the signal distribution, so different types of signals require different crest factors. The sinusoid has the narrowest distribution — sinusoids have a finite peak value, so a crest factor of 3 dB is theoretically enough to avoid all clipping for a sinusoidal signal. In practice, however, noise, distortion, transients, etc. make it wise to leave a few more dB of "headroom" so that overflows will be extremely rare. Modem signals (16 QAM is shown) also tend to have a restricted range, although this would vary with the modulation scheme. A Gaussian signal has a wider distribution and therefore requires a much greater CF to make overflows rare (infinite tails in the distribution make it impossible to eliminate overflows for a Gaussian). Worst of all is the speech signal, which tends to run at low average power except for the occasional peaks that require a large CF.

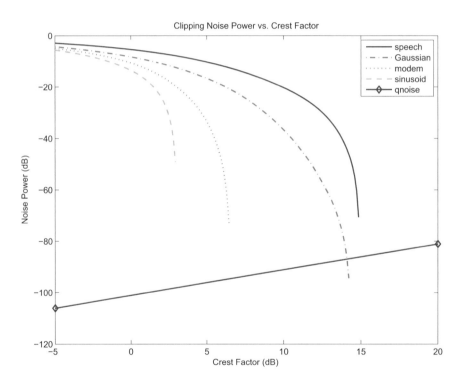

Figure 5.8: A comparison of four signal types in terms of their clipping noise power at various CF values. The qnoise line represents quantization noise for a 16 bit signal format with constant S and B. [26]

Figure 5.8 makes it clear that the signal distribution is important in determining an appropriate signal level, which is always a tradeoff between avoiding clipping and maximizing SNR. Crest factor is important because it limits the dynamic range at the upper end just as the noise floor limits the lower end.

The "qnoise" line in Figure 5.8 shows the quantization noise power associated with format with 16 bits, constant S, and varying Δ (or exponent). Note that the quantization noise power varies with the crest factor because increasing Δ increases both the CF and the noise power. The "qnoise" line is meaningless for the other two situations where noise power is constant, but it provides a reference point for typical 16 bit noise power.

In many systems, crest factor can be set to roughly the same value as the quantization noise power, so that neither dominates. However, in communication systems, clipping can directly limit the bit error rate (BER), while the noise floor is usually limited by the channel, not the system. In this case, the probability of clipping will be driven by the need to preserve the BER.

5.3   COEFFICIENT QUANTIZATION

Particularly in IIR filters, the frequency response can be very sensitive to variations in coefficient values. The filter coefficients in a direct form implementation are simply polynomial coefficients of the transfer function. Small changes in these coefficients due to quantization causes the roots of the numerator and denominator polynomials of the z-transform to move. Changing the position of these roots in z-plane causes the frequency response to change and may even cause the filter to go unstable as poles are moved onto or outside the unit circle.

Similar effects can be observed in the z-transform polynomials of FIR filters, but zeros cannot cause instability, and they do not have the large associated gains that poles do, so that FIR filters suffer less from quantization effects than IIR filters. One exception to this rule of thumb is the depth of the stopband. Only a zero exactly on the unit circle can force an exact zero in the stopband, so zero positioning is critical to obtaining large attenuations in the stopband. Quantizing the coefficients of an FIR filter will misalign the zeros near the unit circle and usually reduce stopband attenuation significantly. One can think of this as a complement of the large gain sensitivity of poles, since zeros have an equivalently large attenuation sensitivity.

Another way to look at coefficient quantization is to consider that with a finite number of coefficient values available, there will be a finite number of possible pole locations. Figure 5.9 shows the pole locations available for four bit coefficients and for six bit coefficients. A number of observations can be made from this figure. First, there is a clear pattern to the pole locations that is consistent with increasing numbers of bits (though the point set will become more dense). Second, the set of complex pole locations is sparse around (not on) the real axis, and particularly near the unit circle at $z = 1$ and $z = -1$. This suggests that very narrow lowpass and highpass filters will be especially difficult to quantize. Third, the pattern of complex poles can be accounted for by observing the relationship between the pole locations in polar coordinates, and the polynomial coefficients. If we write the complex conjugate roots of a second-order polynomial as $z = re^{\pm j\theta}$ and then reduce the transfer function back to polynomial form, we get

$$H(z) \;=\; \frac{1}{1 - a_1 z^{-1} - a_2 z^{-2}} \tag{5.19}$$

$$=\; \frac{1}{\left(1 - re^{\,j\theta} z^{-1}\right)\left(1 - re^{\,-j\theta} z^{-1}\right)} \tag{5.20}$$

$$=\; \frac{1}{1 - 2r\cos(\theta)z^{-1} + r^2 z^{-2}}. \tag{5.21}$$

Dividing the range of a_1 into equal steps is equivalent to dividing the real axis into equal steps, since $r\cos\theta$ is the projection of z onto the real axis. On the other hand, dividing the range of a_2 into equal

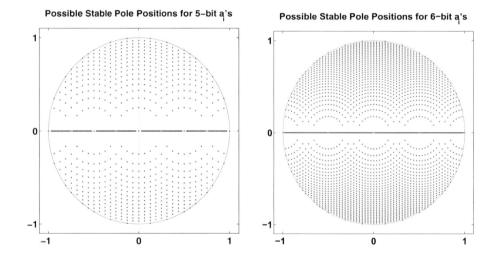

Figure 5.9: All the stable pole locations for a second-order polynomial are shown when the coefficients are represented with five bit, and six bit quantized values.

steps is equivalent to requiring equal steps in $-r^2$, which will result in smaller and smaller steps in r as r approaches 1.

5.3.1   SIGNIFICANT FACTORS IN COEFFICIENT QUANTIZATION
The severity of the quantization effects is effected by several factors.

- **Tightly clustered roots:** When roots are tightly clustered, it is unlikely that the set of available root locations will preserve the relationship between the nearby roots, and often one of the roots will move away from the cluster, changing the frequency response significantly.

- **Roots close to the unit circle:** When a root is very close to the unit circle, it can easily be quantized to a nearby location *outside* the unit circle, causing the filter to become unstable.

Also, roots near the unit circle have strong gain effects on the frequency response, and are, therefore, the frequency response is very sensitive to small changes in pole location.

- **Many roots:** Higher order filters have many coefficients, and each root depends on the value of all the coefficients, so more coefficient changes means a greater potential to change each pole location. Also, the range of polynomial coefficient values increases for higher order filters, and this can make it difficult to maintain precision for small value coefficients if large value coefficients need to be represented in the same Q format.

- **Filter structure:** Direct form filter structures are not the only option available. A number of other structures exist which exhibit better coefficient quantization characteristics than direct form. We will give special attention to the coupled form and cascaded second-order sections form.

- **Roots close to the real axis:** As noted above, for direct form filters, the pattern of available roots is sparse around the real axis, so roots in that region tend to be represented poorly by quantized coefficients.

There are a number of ways to minimize the effects of coefficient quantization and many of these will be discussed in more detail later.

- Use smaller filter sections.

- Select filters with less tightly clustered roots.

- Use more bits of precision.

- Try slight variations on the filter to find which provides the best response under the quantization constraints.

- Scale the coefficients (by choosing the Q format) to advantageously trade-off significant bits and overflow probability.

The rest of this section reviews a popular approach to minimizing coefficient quantization problems: alternative filter structures like the coupled form and cascaded second-order sections. Many other types of alternative structures are good choices for specific applications and are well covered in other sources.

5.3.2   2ND-ORDER COUPLED FORM STRUCTURE

We have discussed the pole pattern imposed by quantizing the direct form coefficients, but the coupled form structure can be used to create a much more regular pattern of pole locations in the z-plane. Figure 5.10 shows the flow graph of the coupled form filter. Note that it requires twice as many multiplies as the equivalent direct form implementation of two poles.

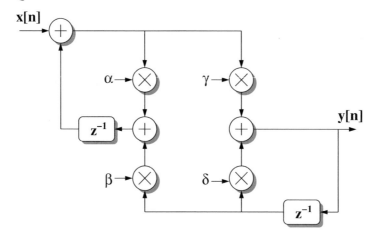

Figure 5.10: The flow graph for a coupled form two pole filter [20].

The transfer function of the coupled form is given by

$$H(z) = \frac{\gamma}{1 - (\alpha + \delta)z^{-1} - (\beta\gamma - \alpha\delta)z^{-2}} \tag{5.22}$$

If $\alpha = \delta = r\cos\theta$ and $\beta = -\gamma = r\sin\theta$ then

$$H(z) = \frac{\gamma}{1 - 2r\cos\theta z^{-1} + r^2 z^{-2}} \tag{5.23}$$

And, comparing with our traditional form,

$$
\begin{aligned}
a_1 &= \alpha + \delta &= 2\alpha &&&= 2r\cos\theta \\
a_2 &= \beta\gamma - \alpha\delta &= -\beta^2 - \alpha^2 &= -r^2(\sin^2\theta + \cos^2\theta) &&= -r^2
\end{aligned} \tag{5.24}
$$

When quantizing α and δ, we get evenly spaced locations along the real axis, just as we did for direct form, since each represents $r\cos\theta$. When we quantize β and γ, we get evenly spaced values along the imaginary axis, as we would expect with values of $r\sin\theta$. This produces a rectangular grid seen in Figure 5.11 instead of the mixed radial pattern produced by direct form quantization.

5.3.3 DIRECT FORM IIR FILTERS - COEFFICIENT QUANTIZATION PROBLEMS

When filter orders are larger than two, additional problems with coefficient quantization appear. The (potentially) large number of taps magnifies the effect of coefficient quantization on the response. The polynomial coefficients cover a wide range of values, requiring either scaling each coefficient separately (near floating point) or losing precision on smaller values.

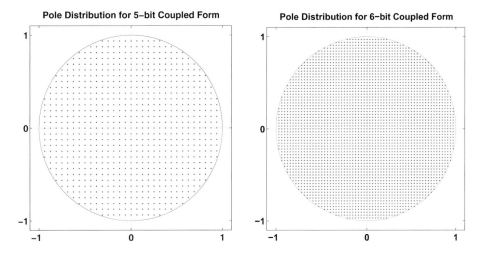

Figure 5.11: For a coupled form filter structure, all the stable pole locations for a second-order polynomial are shown when the coefficients are represented with five bit, and six bit quantized values.

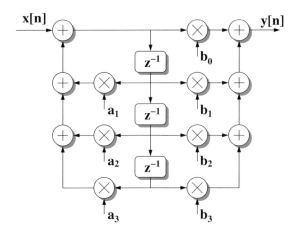

Figure 5.12: The signal flow graph for a fourth order Direct Form II filter implementation.

For a worst-case example of the coefficient range problem, consider a filter with repeated roots at $z = -1$ so that

$$H(z) = \frac{1}{(1 + z^{-1})^k}.$$

As the order increases, the polynomial coefficients follow a binomial sequence (or Pascal's triangle), so that for the kth order example,

$$H(z) = \frac{1}{\displaystyle\sum_{k=0}^{n} \binom{n}{k} z^{-k}}. \tag{5.25}$$

where $\binom{n}{k}$ is defined as

$$\binom{n}{k} = \frac{n!}{k!(n-k)!} \tag{5.26}$$

For even powers, the largest coefficient will always be $\binom{n}{n/2}$ and the smallest coefficient will always be 1. Therefore, as filter order increases, we must be able to accurately represent coefficient values whose ratio grows as nearly 2^n, since[5]

$$\binom{n}{n/2} \approx \sqrt{\frac{2}{\pi}} \frac{2^n}{\sqrt{n}}. \tag{5.27}$$

To demonstrate the effect of increasing filter order on coefficient range, consider the behavior of two filters:

```
>> [b,a] = ellip(2,1,20,0.2)
b =
    0.1432    0.0117    0.1432
a =
    1.0000   -1.2062    0.5406
>> [b,a] = ellip(8,1,20,0.2)
b =
    0.1059   -0.5984    1.6695   -2.9214    3.4966   -2.9214    1.6695
   -0.5984    0.1059
a =
    1.0000   -6.2654   18.2574  -31.9616   36.6063  -28.0235   14.0013
   -4.1801    0.5743
>>
```

In both cases, the coefficients Matlab produces are the coefficients needed to implement the filter in any direct form (recall the a_k's must be negated for use in a signal flow graph). When choosing a Q format for the coefficients, we would like to allow uniform handling of each branch by choosing the same Q format for all the numerator coefficients and the same Q format for all the denominator coefficients. The numerator and denominator are usually handled separately, since their value range is often different.

---

[5]This approximation is thanks to Prof. Kurt Bryan of Rose-Hulman Inst. of Tech.

For the second-order filter, assuming a 16 bit wordlength, the a_k's should be represented in $Q2.14$ since the maximum magnitude is less than 2 but greater than 1. The b_k's can be represented with the standard $Q1.15$, but since none of them is greater than $\frac{1}{4}$, we could also use $Q(-1).17$ to retain more precision. For the eighth order filter, the a_k's require $Q7.9$ to accurately represent a_4 without overflow, leaving just 9 bits to represent a_8. The b_k's require $Q3.13$, and b_0 is small enough that only 10 bits are used.

The point of this example is just to show that under typical constraints, the filter coefficients suffer a great deal more quantization as the direct form structure increases in order. Because of this problem, one might wish to find a way to represent high order filters without using high order direct form structures to reduce the effects of coefficient quantization.

5.3.4   CASCADED SECOND-ORDER SECTION FILTERS

One way to represent a high order filter without high order sections would be to simply cascade the implementation of each of the poles and zeros as first order direct form filters. A cascade of first order sections could be used to implement any high order filter, and since no roots would ever be combined within a section, the coefficients would have a relatively small range. However, many filters have complex roots, and so the first order sections would need to be implemented in complex arithmetic. Complex arithmetic requires many more operations than real arithmetic (for each operation the real and complex parts must be handled separately), so this is a poor choice unless other system constraints require complex signals.

The next best alternative is to use a cascade of second-order sections to build up a filter. With second-order sections, complex conjugate pole pairs and zero pairs can be represented with real coefficients, so no complex arithmetic is needed for real filters, but coefficient range is still relatively limited. A single second-order direct form section can represent two poles and two zeros.

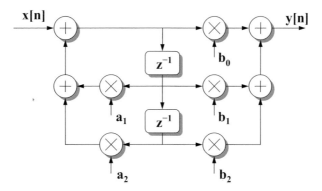

Figure 5.13: A second-order direct form section limits the range needed to represent the coefficients, and they can be cascaded to form high order filters.

We can be certain of the maximum magnitude required of the a_i coefficients because we have already derived the value of the coefficients in polar form. Since

$$H(z) = \frac{1}{1 - 2r \cos(\theta)z^{-1} + r^2 z^{-2}}.$$

and $|\cos(\theta)| \leq 1$ and for stable poles, $|r| < 1$, we know $\begin{cases} |a_1| < 2 \\ |a_2| < 1 \end{cases}$. This implies that no coefficient representation overflow can occur when using $Q2.14$ format for second-order section denominator coefficients. Of course, there is still the extreme case of $a_1 \approx 2$ causing an overflow, and there may be situations where the coefficient values are known to be small enough to allow even more fraction bits, so $Q2.14$ is only a good choice, not the guaranteed best choice.

When second-order sections are cascaded, the system function is just the product of the system functions of each individual section, so high order filters are easily obtained.

Because the more explicit block diagram notation we have been using is too bulky to accommodate multiple sections in a single diagram, we will switch to a more compact flowgraph notation. Beginning with Fig. 5.14, both gains and delays will be shown with a labeled arrow, and adders will occur at intersections with multiple inputs, with a solid circle for emphasis.

$$H(z) = \prod_{k=1}^{N_s} \left(\frac{b_{0,k} + b_{1,k}z^{-1} + b_{2,k}z^{-2}}{1 - a_{1,k}z^{-1} - a_{2,k}z^{-2}} \right) \tag{5.28}$$

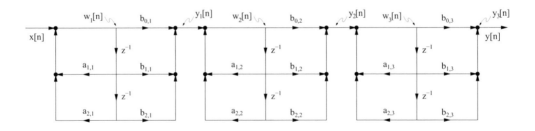

Figure 5.14: A cascade of second-order sections can be used to implement a higher order filter. Intermediate signal names, and an indexing scheme for labeling the coefficients are shown [23].

$$w_k[n] = a_{1k}w_k[n-1] + a_{2k}w_k[n-2] + y_{k-1}[n] \tag{5.29}$$
$$y_k[n] = b_{0k}w_k[n] + b_{1k}w_k[n-1] + b_{2k}w_k[n-2] \tag{5.30}$$

The primary reason for using cascades of second-order sections (SOS) is that the range of required coefficient values is reduced compared to high order direct forms. However, there are other benefits, such as the fact that errors in coefficient values in one section do not effect pole placement in other sections. In direct form, a single badly represented coefficient can change the location of every pole in the entire filter.

In a SOS implementation, there are many possible arrangements of the section orderings and pole pairings which would be equivalent with ideal arithmetic, but produce very different intermediate signal values, and thus different quantization and overflow results. One simple rule of thumb is that within each SOS, the pole and zero pair should be chosen to be as close to each other as possible, so that their net effects will partially cancel and minimize the frequency response changes at the output of each section [13]. The quantization implications of how the sections are ordered will be dealt with later. Keep in mind that an SOS implementation requires more multiplications and adds than direct form, so its advantages are not gained for free.

Note, we could factor out a single gain factor $b_{0,*} = \prod_{k=1}^{N_s} b_{0,k}$ and thus replace each $b_{m,k}$ with $\frac{b_{m,k}}{b_{0,k}}$. The result would require only four multiplies for each section because in every case $b_{0,k}$ would be replaced by 1. However, this is rarely done because we want to retain the ability to scale each section. Using a five coefficient section, we can multiply all the b values of a single section by a constant to adjust the signal amplitude at each section and cancel out the effect with a constant multiplier at the end if desired. Although this ability to scale within the filter is useful, it is also another "free parameter" we must make careful choices about. No such ability to scale the a coefficients is available, because the a_0 coefficient must remain 1.

The process of converting direct form coefficients to SOS coefficients and back is just factoring a polynomial into second-order factors, or multiplying those factors back out.

$$H(z) = \prod_{k=1}^{N_s} \left(\frac{b_{0,k} + b_{1,k} z^{-1} + b_{2,k} z^{-2}}{1 - a_{1,k} z^{-1} - a_{2,k} z^{-2}} \right) \Leftrightarrow \frac{\sum_{k=0}^{M} b_k z^{-k}}{1 - \sum_{k=1}^{N} a_k z^{-k}} \tag{5.31}$$

Although the algebra involved is familiar, factoring fourth order and higher polynomials is nontrivial, and even multiplying out high order polynomials by hand can become tedious and error prone. Fortunately, many tools are available for accelerating these procedures. Matlab provides some basic tools such as `roots()`, `poly()`, and `conv()`.

- **To Convert a System Function Cascade SOS Form**

    1. Factor the polynomials in z (both numerator & denominator)

    2. Combine complex conjugate roots to form second-order polynomials

    3. Check the gain of the result (many tools only identify root locations and may not preserve gain)

    4. Check to make sure $a_{0,k} = 1$ for each section

- **To Convert a Cascaded SOS Form to Direct Form**

    - Multiply the second-order polynomials on top and bottom.

– Alternatively, convolve the coefficients of the numerator polynomials and then of the denominator polynomials.

The following functions are part of the Matlab Signal Processing Toolbox. They allow automated conversion of direct form coefficients into cascaded SOS form, and they also support expression of the filter in terms of the zeros, poles, and the gain. The transfer function form is the direct form coefficients expressed as two vectors, b_i and a_i. The second-order section matrix uses one row to express the six coefficients required for a second-order section, and each row represents a second-order section. Thus, in the notation of Figure 5.14, the SOS matrix has the form

$$\text{SOS matrix} = \begin{bmatrix} b_{0,1} & b_{1,1} & b_{2,1} & a_{0,1} & a_{1,1} & a_{2,1} \\ b_{0,2} & b_{1,2} & b_{2,2} & a_{0,2} & a_{1,2} & a_{2,2} \end{bmatrix}. \tag{5.32}$$

Of course $a_{0,k} = 1$ for every section, so the matrix represents a five multiplier section with each row.

tf2sos Transfer Function to Second-Order Section conversion. *This function has options for sorting and scaling the gains in the SOSs to improve numerical performance.*

zp2sos Zero-pole-gain to second-order sections model conversion. *This function has the same options as tf2sos.*

sos2zp Second-order sections to zero-pole-gain model conversion.

sos2tf 2nd-order sections to transfer function model conversion.

Quantized SOS Filter Example — 12th Order Elliptic The effects of coefficient quantization, and the benefits of cascaded SOS form are most dramatic in high order filters. Consider the 12th order bandpass elliptic filter [23] whose frequency response is shown in Figure 5.15. The pole zero plot for the same unquantized filter is shown in Figure 5.16. The filter can be designed in Matlab using the command [b,a] = ellip(6,0.1,,40,[0.32 0.41]). If we constrain all the b_i coefficients to the same Q format representation, and likewise for the a_i coefficients, the resulting coefficients and their quantized values are shown in Table 5.3.

We can see that for the b_i coefficients, the ratio between the largest and smallest magnitude is greater than 67, which means that more than 6 bits are lost to shifting the radix point in representing the smallest value. With the a_i coefficients, the ratio is greater than 117, requiring almost 7 bits to be zero in the smallest value. Because the precision is reduced, and the high order filter is more sensitive to coefficient errors, the result is that the implementation would fail because the poles are moved outside the unit circle, and the quantized filter is unstable. Figure 5.17 shows what the frequency response of the quantized filter would be if it were stable, and Figure 5.19 shows the pole zero plot of the quantized filter. With poles both inside and outside the unit circle, the quantized filter cannot be stable whether operating as a causal system, or as an anti-causal (time-reversed) system.

Fortunately, we can implement the filter using a cascaded SOS form, and the results are much better. Figure 5.19 shows the frequency response of the SOS form filter quantized to 16 bit

	unquantized	quantized
Table 5.3: Comparison of coefficients in a direct form filter.		
b_i **Q format**		$Q1.15$
b_0	0.010402294826284	0.010406494140625
b_1	−0.046981738177133	−0.046966552734375
b_2	0.140584418862936	0.140594482421875
b_3	−0.291587543984708	−0.291595458984375
b_4	0.483154936264528	0.483154296875000
b_5	−0.638421915199625	−0.638427734375000
b_6	0.705123004845762	0.705108642578125
b_7	−0.638421915199625	−0.638427734375000
b_8	0.483154936264528	0.483154296875000
b_9	−0.291587543984709	−0.291595458984375
b_{10}	0.140584418862936	0.140594482421875
b_{11}	−0.046981738177133	−0.046966552734375
b_{12}	0.010402294826284	0.010406494140625
a_i **Q format**		$Q8.8$
a_0	1.000000000000000	1.000000000000000
a_1	−4.736891331870115	−4.738281250000000
a_2	14.770074008149495	14.769531250000000
a_3	−31.339008666596580	−31.339843750000000
a_4	52.290097782457892	52.289062500000000
a_5	−68.277884768136261	−68.277343750000000
a_6	73.209425975401615	73.210937500000000
a_7	−63.111029848303829	−63.109375000000000
a_8	44.673344222036803	44.671875000000000
a_9	−24.741797451330299	−24.742187500000000
a_{10}	10.774944914870451	10.773437500000000
a_{11}	−3.191627411828395	−3.191406250000000
a_{12}	0.622743578239240	0.621093750000000

coefficients. The pole zero plot is not shown because the differences between the quantized and unquantized root locations are difficult to see. With SOS form, the resulting frequency response in Figure 5.19 is very close to the ideal response shown in Figure 5.15.

As described before, the filter's gain can potentially be adjusted by scaling the $b_{i,k}$ coefficients in each section. To avoid ambiguity in this example, all the gain has been factored out into a separate gain term (also quantized to 16 bits), and thus all the $b_{0,k}$ coefficients are exactly 1, along with

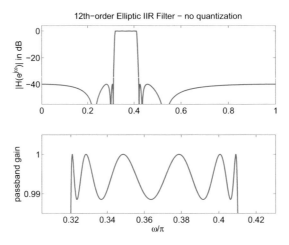

Figure 5.15: The frequency response of a 12th order bandpass filter of elliptic design. A linear plot of the passband gain is shown to emphasize the equiripple behavior [23].

the $a_{0,k}$ coefficients. The unquantized gain value is 0.010402294826284, and the unquantized SOS matrix has the values shown in Table 5.4.

		Table 5.4:	Unquantized SOS matrix.		
$b_{0,k}$	$b_{1,k}$	$b_{2,k}$	$a_{0,k}$	$a_{1,k}$	$a_{2,k}$
1	0.195430137972198	1.000000000000022	1	−0.664513985433423	0.852588727671713
1	−1.508636017183002	0.999999999999984	1	−0.873355320687181	0.860012128095767
1	−0.412640684596503	0.999999999999658	1	−0.564681165614170	0.932247834436929
1	−1.177232208175121	0.999999999999855	1	−1.019134004225044	0.939497815061180
1	−0.492675670575463	1.000000000000334	1	−0.543644167598933	0.983728242806233
1	−1.120723792458177	1.000000000000142	1	−1.071562688311368	0.985740948265446

The quantized gain value is represented as $Q(-5).21$ and has a value of 0.010402202606201, which is optimal for a 16 bit word, since it has its own separate radix shift just as if it were in floating point. Because the maximum magnitudes of both the $b_{i,k}$ coefficients and the $a_{i,k}$ coefficients fall between 1 and 2, all the elements of the SOS matrix are quantized using $Q2.14$. The resulting values are given in Table 5.5.

Keep in mind that in a specific implementation, design choices would be made about how to distribute the gain among the sections (usually by scaling the $b_{i,k}$ coefficients in each section), and the $b_{i,k}$ coefficients would then have a somewhat larger range of values and perhaps a different Q format than the $a_{i,k}$ coefficients.

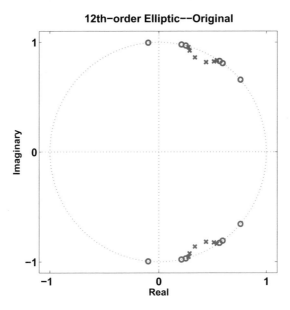

Figure 5.16: In the original filter, the poles are clustered near the unit circle with zeros nearby to create a sharp transition band [23].

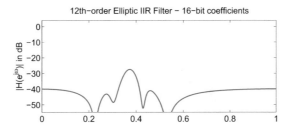

Figure 5.17: After representing the filter transfer function with 16 bit coefficients, its value along the unit circle is no longer similar to the desired bandpass specification. In fact, it is not technically a frequency response, since the filter has become unstable [23].

MATLAB Filter Design and Analysis Tool (fdatool) Demo   For readers who have installed Matlab's Filter Design Toolbox and Fixed-Point Toolbox, the `fdatool` graphical user interface (GUI) can be used to demonstrate the same concept interactively. Simply follow the steps below.

1. Launch `fdatool` by typing `fdatool` at the command prompt.

2. Under "Response Type" choose "Bandpass."

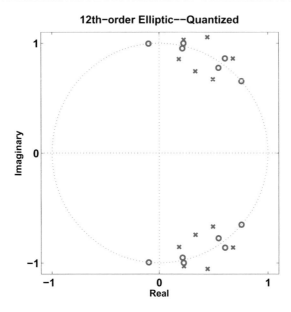

Figure 5.18: Quantizing the polynomial coefficients in direct form caused the poles and zeros to relocate. The movement of the poles outside the unit circle results in an unstable filter [23].

Table 5.5: Quantized SOS matrix.					
$b_{0,k}$	$b_{1,k}$	$b_{2,k}$	$a_{0,k}$	$a_{1,k}$	$a_{2,k}$
1	0.195434570312500	1.000000000000000	1	−0.664489746093750	0.852600097656250
1	−1.508605957031250	1.000000000000000	1	−0.873352050781250	0.859985351562500
1	−0.412658691406250	1.000000000000000	1	−0.564697265625000	0.932250976562500
1	−1.177246093750000	1.000000000000000	1	−1.019104003906250	0.939514160156250
1	−0.492675781250000	1.000000000000000	1	−0.543640136718750	0.983703613281250
1	−1.120727539062500	1.000000000000000	1	−1.071533203125000	0.985717773437500

3. Under "Design Method" choose "IIR Elliptic."

4. Leave all other fields at their default values. Click the "Design Filter" button.

5. You should see a log plot of the bandpass Magnitude Response of a 10th order SOS form filter with floating point coefficients.

6. Convert the filter to a single direct form section by selecting the "Edit⇒Convert to Single Section" from the pull down menu at the top of the window.

7. Now select the "Set quantization parameters" tab from the icons along the lower left edge of the window (the icon looks like a quantizer's stair step diagram). *When the fdatool window is active, mousing over the tabs should produce a tooltip label for each tab.*

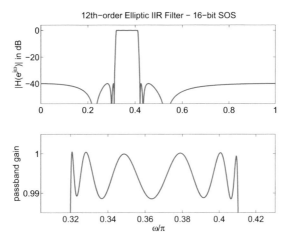

Figure 5.19: Quantizing the coefficients in cascaded SOS form results in a much more accurate filter response. Small variations are visible in the linear plot of the passband [23].

8. Set the filter arithmetic option to "Fixed–point". You should see the frequency response curve change to a poor approximation of the bandpass filter, and the "Stable" field in the current filter information block change to "No" indicating that the quantized filter is not stable.

9. Choose the Pole/zero plot button from the icons across the top of the window to see a comparison between the poles and zeros of the quantized and unquantized filters. You may also wish to apply the zoom tool for a better view.

10. To return your design to SOS form, choose "Edit⇒Convert to Second-Order Sections" from the pull down menu, and then choose the design filter tab from the lower left icons, and click the Design Filter button again.

11. Experiment with different filter parameters to see how sensitive various filter designs are to coefficient quantization errors.

CHAPTER 6

Quantization Effects – Round-Off Noise and Overflow

6.1 ROUND-OFF NOISE: PRODUCT AND SUM QUANTIZATION

A major problem with fixed-point arithmetic is that every multiply dramatically increases the need for precision to represent the ideal result. For a quick example of this effect, consider the product of $\frac{1}{8} \times \frac{1}{8} = \frac{1}{64}$ in $Q1.3$ arithmetic. Recall from Chapter 4 that $\frac{1}{8}$ is the smallest representable number in $Q1.3$, and that the product will naturally produce a $Q2.6$ number which can represent the result, but which must be truncated or rounded in order to return to $Q1.3$ for further operations. Figure 6.1 is a graphical representation of a similar example in $Q1.7$ form. The only alternative to reducing the precision during processing is allowing the precision bits to grow at each operation, which rapidly becomes impractical. The usual solution is to maintain higher precision results as long as possible in order to minimize the number of times results are rounded or truncated during processing.[1]

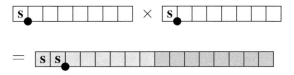

Figure 6.1: The product of two $Q1.7$ numbers naturally produce a 16 bit result in $Q2.14$ format. Often we will need to reduce the result back to an 8 bit format for further operations. The appropriate format of the 8 bit result will depend on the algorithm, the data, overflow requirements, and the system architecture.

When using higher precision formats, even more bits are needed to represent the ideal result, and more are lost if the result is returned to the original format. For large numbers of bits, these reductions in precision behave almost identically to a quantization step from a continuous valued (analog) signal to a discrete-valued signal (like $Q1.15$). Therefore, modeling the quantization step as the injection of noise in a linear model is just as useful in analyzing round-off error (now considered round-off noise) as it was in the previous consideration of signal quantization. The only difference is that the noise is now injected within the filter system so that we must determine the effect of whatever portion of the filter the noise passes through on its way to the system output, and there will often be multiple quantization points (noise sources) within a filter or system.

---

[1] An interesting counter example occurs with CIC filters where the results may begin at low precision and must be allowed to grow in precision in multiple stages. Lyons discusses this procedure [17].

The fact that processing introduces additional noise to a signal may seem surprising to readers who have been well indoctrinated on the "perfection" of digital signals and audio equipment. However, it is well known that the source for 16 bit audio formats like the CD has usually been recorded using 24 bit or higher professional equipment. One excellent reason for recording the extra precision is that noise introduced by processing must corrupt the lowest bits of the signal. If, during processing, the extra bits allow much smaller quantization steps than the much larger quantization step introduced when producing the CD format, then some processing can occur without reducing the quality of the final 16 bit product. If processing occurred directly on 16 bit data, then the 16 bit result might be noticeably more noisy because of the larger round-off errors inherent in the processing.

Adders also produce outputs with more bits than the inputs, but the effect is much less severe than with multipliers. Consider the example of adding $-1 + (-1) = -2$ in $Q1.3$ format. It is clear that the quantity -2 cannot be represented in $Q1.3$ and so the algorithm designer may wish to use $Q2.3$ format for the result in order to avoid overflow. If hardware constraints mean that a 5 bit word is impractical, then the result my need to be stored as $Q2.2$, but this requires the loss of the least significant bit, and this is a quantization.

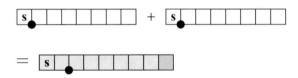

Figure 6.2: When adding two values in the same Q format, the result could be twice as large as the operands, so to prevent overflow, another bit is needed. If the architecture limits the word-length, that extra bit may require the loss of a less significant bit.

The usefulness of the uniform additive white noise approximation is limited in this case, because the distribution of the quantization error can be thought of as Bernoulli distributed[2] (the result of the quantization is always either exactly unchanged, or off by $\Delta/2$. This is rather poorly approximated by the uniform distribution. However, sums of many repeated single bit quantization errors can be usefully approximated as Gaussian distributed values, even though they are actually discrete (approximately[3] binomial) in distribution.

Fortunately, both of these problems can be minimized or avoided entirely by certain hardware provisions that have proven quite popular in DSP architectures.

- Multiply round-off noise can be minimized by providing for double precision sums (accumulator register) so that, for example, the many products resulting from an FIR filter computation

---

[2]This assumes truncation or simple rounding, round-to-even gives three possible error values.
[3]The distribution is approximate because we can't guarantee that the error samples are independent, but we assume that they are.

can be added as double precision values and the output produced with a single quantization step instead of one for each product.

- Addition round-off to avoid overflow can be
  - avoided by providing extra "guard" bits in the accumulator register so that overflows cannot occur unless the sum is larger than the guard bits can represent, or
  - eliminated for a filter by scaling the data or coefficients so that the sum cannot attain values that could overflow.

The use of a double precision accumulator can dramatically reduce the number of quantization operations required, since data can be kept in double precision until some non-addition operation is required, such as another multiply, or storage to memory. Guard bits have become standard on many DSP μP architectures largely because of their assumed presence in several popular speech coding standards. The convenience of implementing speech coding standards is a factor in the popularity of the architecture using 16 bit register files, and 40 bit (32 bits for double precision products, plus 8 "guard" bits) accumulator registers. Fortunately, more than doubling the bit length of the adder hardware increases chip complexity far less than doubling the bit length of the multiplier hardware would have. Scaling the data or filter coefficients to avoid overflow is a powerful approach, but involves signal level tradeoffs that are roughly equivalent to the quantization operations required above. Scaling tradeoffs will be discussed later along with other scaling issues.

Figure 6.3 shows how the noise injected in a particular filter flow-graph can be considered to be added to the output of each multiplier and adder. At first glance, it appears that each noise input will be processed by a different partial filter, but in fact, we only need to consider two filter responses to accurately analyze this system.

To determine the behavior of the noise at the output of the filter due to noise injected at a certain node, we need to determine the impulse response from that node to the output. Imagine that rather than injecting noise as a signal at $u_1[n]$ we instead injected an impulse. We can denote[4] the impulse response from here to the output as $h_{u_1,out}[n]$.

If we calculate the impulse response $h_{u_1,out}[n]$ we will find that it is exactly the same as $h_{x,out}[n]$ and further,

$$h_{x,out}[n] = h_{u_1,out}[n] = h_{u_2,out}[n] = h_{v_0,out}[n] = h_{v_1,out}[n]. \tag{6.1}$$

This works because *moving an added input across a sum has no effect.* The same principle means that $u_3[n]$, $u_4[n]$, $u_5[n]$, $v_2[n]$, and $v_3[n]$ are all effectively at the output so that $h_{u_5,out} = \delta[n]$ as well as the other nodes at the output. In other words, the noise injected at the output will not be filtered at all, and the noise injected at the input will be filtered by the entire filter.

If we perform a similar analysis on a Direct Form II transpose filter, we will need to know that under our assumptions that the noise sources are stationary and uncorrelated, we can ignore

---

[4]We choose this slightly cumbersome notation because later we will be interested in impulse responses from an impulse at the input to the response at a certain node, so we want to clearly distinguish these two cases.

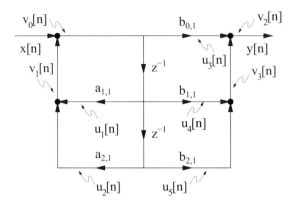

Figure 6.3: A flow graph of a second order Direct Form II filter showing where noise sources would be injected assuming quantization operations were required after every multiply and add. The u_i noise sources represent product round-off noise, and the v_i sources represent sum round-off noise.

the effects of shifting an input across a delay element as well, which will simplify analysis for several cases. Shifting an input across a constant multiply is allowable if the change in gain is taken into account. Possibly most important, though, is the fact that we cannot generally shift an input across a branch (where a signal is split into two or more paths).

A key piece of intuition in analyzing flow-graphs is that any signal that reaches the feedback path will be filtered by the feedback portion of the filter (the poles). Signals that are injected after the feedback path may be filtered (or not) by the rest of the filter, but the poles will have no effect on them.

Once we have determined the effect of the partial filter between where a quantization noise signal is injected and the output, we can determine the resulting noise power, and the power spectrum associated with each noise source. We can also consider combining the effects of the multiple noise sources.

- To find noise power at the filter output, we must account for the effect of the (partial) filter on its power. See the derivation of (3.50), the result of which is repeated here for convenience.

$$\sigma_{out}^2 = \sigma_e^2 \sum_{n=-\infty}^{\infty} \left| h_{eff}[n] \right|^2 \tag{6.2}$$

- This must be computed for every noise source.

- Noise sources are assumed to be independent so their powers add.

- Noise sources are assumed to be spectrally white.

For the specific example of Figure 6.3, we may wish to assume that all the product quantization noise sources $u_i[n]$ have a power of σ_e^2, and that the input signal is scaled such that no sum quantization $v_i[n]$ is necessary. Then we find that the noise power at the output of the filter due to $u_1[n]$ is

$$\sigma_{out(u_1)}^2 = \sigma_e^2 \sum_{n=-\infty}^{\infty} \left| h_{u_1,out}[n] \right|^2 \tag{6.3}$$

and the noise power at the output of the filter due to $u_3[n]$ is

$$\sigma_{out(u_3)}^2 = \sigma_e^2. \tag{6.4}$$

Since the effects of noise sources at the filter input are equal and additive, and the same is true of the noise sources at the output, the total noise power at the filter output for this example would be

$$\sigma_{out(tot)}^2 = 2\sigma_e^2 \left(\sum_{n=-\infty}^{\infty} \left| h_{u_1,out}[n] \right|^2 \right) + 3\sigma_e^2. \tag{6.5}$$

One way to interpret this result would be to think of two noise sources (at the filter input) filtered by the entire filter, plus three noise sources (at the filter output) filtered by nothing.

A similar formulation allows us to determine the power spectrum of the noise at the filter output. We begin by assuming that each noise source has a power spectrum of $P_{out}(e^{j\omega})$ and that the individual sources are not correlated with each other, so that their spectra can be simply added. We have already established that filtering a random process modifies the power spectrum to produce an output spectrum of

$$P_f(e^{j\omega}) = |H(e^{j\omega})|^2 P_e(e^{j\omega}). \tag{6.6}$$

With these assumptions, we can easily establish that the spectrum of the noise at the filter's output is determined by combining the effects of all the noise sources, so that

$$P_{out(tot)}(e^{j\omega}) = 2|H_{u_1,out}(e^{j\omega})|^2 P_e(e^{j\omega}) + 3P_e(e^{j\omega}). \tag{6.7}$$

Since we are also assuming that quantization noise is white, we know that the spectrum is constant, and therefore $P_e(e^{j\omega}) = \sigma_e^2$. Now, (6.7) reduces to

$$P_{out(tot)} = \left(2|H_{u_1,out}(e^{j\omega})|^2 + 3 \right) \sigma_e^2. \tag{6.8}$$

6.1.1 CALCULATION EXAMPLE

Here we will consider a procedure for calculating the round-off noise power and spectrum based on a filter as it might be implemented in a typical DSP μP. We will also describe a method for simulating the operation of the filter using Matlab's Fixed Point Toolbox so that the round-off noise signal can be isolated from the ideal[5] output signal.

---

[5] Here, "ideal" is defined as the double precision floating–point result because it is far better precision than 16 bit fixed point. Clearly, double precision floating point is not truly ideal precision.

In a DSP μP, the architecture has fixed length register and bus sizes, so processing values in the natural word-length of the architecture is far more efficient than any other word-length. Although many different architectures are available in the market, the assumptions below are meant to represent a fairly common set of design choices.

- In many common DSP architectures, efficient processing and common practice lead to the assumptions that

  - arguments of a multiply are 16 bit,
  - values that must be stored, such as delays and outputs, are 16 bit.

- Sums are usually stored in a double length (32 bit or 40 bit) accumulator.

- Feedback coefficients for 2nd order sections have a range of ± 2 and may use $Q2.14$.

- Scaling operations may require additional quantizations.

- If there is no scaling between sections, values may be passed as 32 bit data ($Q1.31$).

Given these assumptions, many practical designs are possible, and the best choice for signal formats may depend on many constraints, including signal level and data behavior, the cost of overflow events, the provisions in the architecture for shifting product results, and many others. In Figure 6.4, we show a generic set of format choices which may not be appropriate for any given design specifications, but will serve as a representative design in this example.

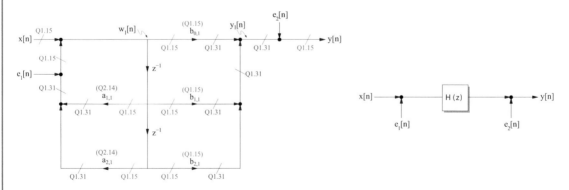

Figure 6.4: A Direct Form II flow-graph with signal formats designated on each signal path, and on the filter coefficients. Also shown are the noise sources associated with the quantization process, both in a detailed diagram, and in an equivalent block diagram.

If we consider $y[n]$ the output of the filter in the z-domain, we can write it in terms of the signals that are filtered and those that are not, so that

$$Y(z) = (X(z) + E_1(z))H(z) + E_2(z). \tag{6.9}$$

However, we can also consider the results in terms of those associated with the

$$\text{signal} = X(z)H(z), \text{ and the} \tag{6.10}$$
$$\text{round-off noise} = E_1(z)H(z) + E_2(z). \tag{6.11}$$

As before, when noise signals are *uncorrelated*, the noise power (and the power spectra) simply add, so we can obtain

$$\sigma_y^2 = \sigma_1^2 \sum_n h^2[n] + \sigma_2^2. \tag{6.12}$$

Figure 6.5 shows how we can calculate the round-off noise referred to in (6.11). In a typical simulation environment, and specifically in Matlab's Fixed Point Toolbox, it is possible to calculate filter results in floating–point, or in fixed–point, but not both simultaneously, at least not automatically. Therefore, we must "manufacture" any comparisons we are interested in by computing both the floating–point and the fixed–point results. This is most conveniently done as part of a single function such as the `filterboth()` function used by the authors.[6] It is very important to observe that if we wish to isolate the round-off noise, the floating–point filter must operate using the fixed–point coefficients, or the round-off noise will be combined with the frequency response differences associated with coefficient quantization.

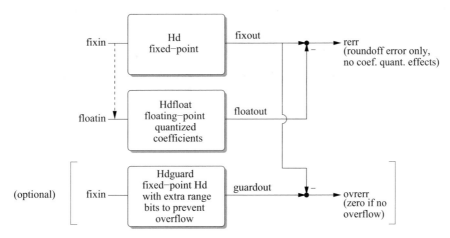

Figure 6.5: A block diagram of a method for computing the round-off error signal in a simulation environment. Provisions are also shown for detecting overflow.

If needed, a similar approach can be used to detect overflow. In this case, the filter's output is computed using extra bits (added to the MSB end of the data word for extra range instead of to the LSB end which would give extra precision). If the two fixed–point results return identical results, no overflow has occurred, but if the results differ, at least one overflow must have occurred. By simply

[6] See the website for the source file.

comparing the outputs, there is no way to determine how many overflow events occurred because all subsequent values are likely to differ. There is also no way to use the output to detect overflows that were subsequently corrected, as can occur with wrapping overflow. If the simulation environment supports overflow logging, overflows could be detected as they occur, and not just if they effect the output, and then output comparison would likely be inefficient and unnecessary.

6.2 OVERFLOW AND SCALING

The problem of overflow can generally be dealt with by scaling the signal so that overflow events occur rarely enough to satisfy the system's performance specifications. Probability of overflow has been discussed above in Section 5.2.5. The previous discussion assumed that the peak signal level was under our control through scaling, but in a filter operation, it isn't always obvious how the intermediate signal results are being scaled, so in this section we will be analyzing filters to determine appropriate scaling values so that overflow in the intermediate values will be avoided.

6.2.1 OVERFLOW REVIEW

First, recall some of the basic properties of overflow in fixed–point representation.

- Using $Q1.15$ data, any value anywhere in the system which reaches a value $|x[n]| > 1$ causes an overflow.[7]

- A natural multiplication of two $Q1.15$ numbers results in a $Q2.30$ number which can never overflow. This is true for the natural results of all multiplies, but if in the $Q1.15$ product example we wish to reduce the result back to $Q1.x$, there remains the unique pathological case of $-1 \times -1 = 1$ which can overflow.

- Addition is much more likely to cause an overflow when attempting to retain the same Q format.

$$
\begin{array}{rl}
0.110 & = 6/8 \\
+0.010 & = 2/8 \\
\hline
1.000 & = -1 \quad \text{not } 8/8!
\end{array}
$$

- Some processors, systems, or representations can allow "guard bits" to handle overflows gracefully. This approach can be more efficient, since overflow checking need not be done after every addition, and more sophisticated overflow handling may be practical.

- Saturation is the process of detecting overflow, and forcing the result to the *maximum possible* value with the same sign.

    - Reduces error compared to wrapping

---

[7]This is a slight oversimplification since we know that two's complement notation is not symmetric. We will often approximate the range for $Q1.x$ data as ± 1 for convenience. Fortunately, for reasonable numbers of bits, a correction factor of $(1 - \Delta/2)/1$ is negligible.

– Requires more hardware or instructions

– Can't recover from "temporary" overflows as guard bits can

6.2.2   SCALING

A typical approach to preventing overflow in a filter or other system is to multiply the input signal by a small constant to scale down the values and therefore reduce any intermediate signal peaks which might cause an overflow inside the system. Of course, in fixed–point, we are limited not only by a maximum value but also by the minimum value. We model the signal quantization as a noise source, and the noise power is set by the Q format, not by the signal level. Scaling down the signal generally reduces the signal power, and since noise power remains fixed, SNR is reduced. It is even possible to scale the signal down so much that all the information is lost to underflow, so there are definite limits to the amount of scaling that is practical.

As a general rule of thumb, we wish to maximize the signal level (without overflow) wherever possible so that the signal will dominate the quantization noise. There will always be a trade-off between avoiding overflow and maximizing SNR.

A cascaded SOS IIR filter is an excellent example of how scaling can be applied, both because it is commonly used, and because the signal tends to be both amplified and attenuated in each stage, or section. Thus, we want to carefully scale the signal at each stage to maximize SNR and minimize overflow.

6.2.3   NORMS

To avoid overflow, we require useful measurements of the signal's variation as it passes through the system. We can avoid overflow entirely using one measure, or simply avoid overflow of sinusoids using another. A great deal of theoretical foundation that we can depend on exists for a general class of vector measurements called norms.

For our purposes, a norm is a just measure of vector length. The L_p norm, defined for a length N sequence as

$$||h[n]||_p = \left(\sum_{n=-\infty}^{\infty} |h[n]|^p \right)^{1/p}. \tag{6.13}$$

For scaling purposes, we will be interested in applying these norms to impulse responses (time–domain), and frequency responses (frequency–domain). For clarity, we will denote the L_p norm in the time–domain as L_{pt}, and in the frequency–domain as $L_{p\omega}$. We can express the $L_{p\omega}$ norm as

$$||H(e^{j\omega})||_p = \left(\frac{1}{2\pi} \int_0^{2\pi} \left| H(e^{j\omega}) \right|^p d\omega \right)^{1/p}. \tag{6.14}$$

Since the L_2 norm is also an energy measure, we can apply Parseval's relation [23] as

$$||h[n]||_2 = ||H(e^{j\omega})||_2 \tag{6.15}$$

and there is therefore no ambiguity between L_{2t} and $L_{2\omega}$.

Since it is more practical to compute the discrete version of the Fourier transform, when applying the $L_{p\omega}$ norm we will apply it to the discrete sequence (the DFT)

$$H[k] = \sum_{n=0}^{N-1} h[n]e^{-j2\pi nk/K} \text{ for } k = 0..K-1, \tag{6.16}$$

where $\omega_k = 2\pi k/K$. The result is the approximation

$$||H(e^{j\omega})||_p \approx \left(\frac{1}{K} \sum_{k=0}^{K-1} |H[k]|^p \right)^{1/p}. \tag{6.17}$$

The discrete approximation will be more accurate as K increases, with equality as $K \to \infty$. The value of K is intentionally left ambiguous, since it depends on the need for accuracy. It is useful to note that an FFT may be effective when $K = N$, and the definition allows zero padding of $h[n]$ when $K > N$, but an FFT is wasteful if $K < N$.

For the purposes of scaling, it is extremely useful to order the norms. It can be shown that [13]

$$||h[n]||_1 \ge ||H(e^{j\omega})||_\infty \ge ||h[n]||_2 = ||H(e^{j\omega})||_2 \ge ||H(e^{j\omega})||_1 \ge ||h[n]||_\infty. \tag{6.18}$$

Although all these norms are occasionally used, $||H(e^{j\omega})||_\infty$ is most common, and we will only discuss the three most conservative choices: $||h[n]||_1$, $||H(e^{j\omega})||_\infty$, and $||h[n]||_2$. We may also refer to these as L_{1t}, $L_{\infty\omega}$, and L_2, respectively.

The computation of the time–domain norms should be straightforward, and the computation of $||H(e^{j\omega})||_\infty$ can be reduced to

$$||H(e^{j\omega})||_\infty \approx \left(\frac{1}{K} \sum_{k=0}^{K-1} |H[k]|^\infty \right)^{1/\infty} = \max_k |H[k]| \tag{6.19}$$

because only the largest term dominates, and the factor $\frac{1}{K}$ is raised to the $\frac{1}{\infty}$ power. Of course, since $|H[k]|$ merely samples $|H(e^{j\omega})|$, a sharp peak in magnitude could be missed by the sampling. Therefore, the approximation is best when K is large.

6.2.4 L_{1t} SCALING

First we will consider the mathematical problem of eliminating all overflow in a filter, and then later we will examine some measures to improve efficiency. Overflow could occur at any node in the filter, so we will denote the signal at the rth node as $x_r[n]$. The system input node will be $x_0[n]$.

If we assume signals in the system do not overflow when $|x_r[n]| < 1$, then we need to ensure that the peak value at every node has a magnitude less than 1. Now consider that every node in the (linear) system has an impulse response $h_{0,r}[n]$. The input $\pm\delta[n]$ is the largest value of input peak that can be present at the input. Given the system input, we can easily compute the intermediate signal at the rth node using convolution. Then

$$x_r[n] = \sum_{m=-\infty}^{\infty} x_0[n-m]h_{0,r}[m] \tag{6.20}$$

and we can produce the maximum possible value of $x_r[n]$ at $n = 0$ by choosing the worst case $x_0[n]$ for node r to be

$$x_0^{<wc_r>}[m] = \text{sign}(h_r[-m]). \tag{6.21}$$

This is simply choosing the system input so that in each summation, the largest possible value, $|h_{0,r}[m]|$ is added to the total, so that

$$\max_n(x_0^{<wc_r>}[n]) = x_r^{<wc_r>}[0] = \sum_{n=-\infty}^{\infty} |h_{0,r}[n]|. \tag{6.22}$$

We have just shown that at node r, the maximum possible value that may appear is $\sum_{n=-\infty}^{\infty} |h_{0,r}[n]|$.

It should be clear that no overflow can occur at node r if the input signal amplitude is reduced to $s_r x_0[n]$, where

$$s_r = \frac{1}{\sum_{n=-\infty}^{\infty} |h_{0,r}[n]|}, \tag{6.23}$$

and no overflow can occur at any node if the input signal amplitude is reduced to $s x_0[n]$ where

$$s = \frac{1}{\max_r \sum_{n=-\infty}^{\infty} |h_{0,r}[n]|}. \tag{6.24}$$

The scaling method described above guarantees no overflow at any node, and is sometimes referred to as absolute[8] scaling.

Now that we have established that it is possible to avoid all overflow, we should point out that L_{1t} scaling is not commonly used because in many applications, an occasional overflow event is acceptable, but the low SNR required to avoid all overflow is not. Consider the form of the "worst-case" signals described in (6.21). Such a pathological signal input is quite unlikely in most systems, and a system which provides for such absolute worst-case conditions is giving up precision at the low end of the signal range to do so. This extremely conservative approach only makes sense where overflow is completely intolerable. For example, one might imagine that nuclear missile guidance systems would need to eliminate overflow in every case. Another example might be a communication system with a requirement for an extremely low bit error rate (BER). Since an overflow would be likely to cause a bit error, overflows would need to be made correspondingly rare, although this

---

[8]Absolute here can be thought of as referring to the sum of absolute values, or to the fact that absolutely no overflow is allowed, but probably the former is the most precise meaning.

example may not require elimination of every possible overflow. On the other hand, an audio system like an MP3 player may tolerate as many as four or five overflow events at the output in a single 3 minute song. Overflow events are much more tolerable in audio when they are handled with saturation than with wrapping, since saturation usually avoids the audible "click."

6.2.5 $L_{\infty\omega}$ SCALING

A sinusoidal signal would be far more likely to appear in most systems than the "worst–case" signals discussed above, and therefore another sensible approach to scaling would be to prevent overflows for any pure sinusoidal signal. This can be accomplished by ensuring that frequency domain transfer function from the system input to each node never exceeds a gain of 1. Under this constraint, a full scale sinusoid at the input can never be amplified to a larger amplitude, so no overflow can occur for sinusoidal signals. If we guarantee this condition by scaling the input signal down, the required scale factor will be

$$s = \frac{1}{\max_r ||H_{0,r}(e^{j\omega})||_\infty} \, , \tag{6.25}$$

but we will usually compute s with a discrete approximation to $H_{0,r}(e^{j\omega})$. Keep in mind that truly pure sinusoids last forever, so there is no provision for starting and ending transients, nor is there provision for harmonics, etc. but on the other hand, there is no requirement to operate the system at exactly full scale. A practical design choice may be to scale down a little more than what is required by the $L_{\infty\omega}$ scaling, but since this is a less conservative method than L_{1t} scaling, the resulting SNR is still likely to be better than for the L_{1t} case. Unless the system's input signals are actually sinusoidal, testing with real data will be required to ensure that the probability of overflow is acceptable.

The assumption of sinusoidal data is especially applicable in systems where narrow-band signals are typical[9]. Another reason the $L_{\infty\omega}$ scaling method can be useful is because it is relatively easy to test the system for overflow behavior with swept sinusoidal inputs. Because it is a reasonable compromise between overflow avoidance and SNR, and because its underlying assumptions apply to many actual systems, $L_{\infty\omega}$ scaling is commonly used, and is often a good starting point in an effort to appropriately scale a system.

Because computing with continuous functions is difficult, the computing of (6.25) is often accomplished by simply sampling the frequency range at discrete points, producing the approximation

$$s \approx \frac{1}{\max_{r,k} |H_{0,r}(e^{j\omega_k})|} \, . \tag{6.26}$$

If the frequency response is smooth, this approximation is well-behaved, but in the case that a pole appears near the unit circle, the resulting sharp peak in frequency response must be sampled very finely to obtain a good estimate of the peak height, and therefore the approximation is sensitive to the frequency sampling grid. Fortunately, very narrow-band peaks must be driven at exactly the resonant frequency for a long period of time to integrate the input long enough to produce a peak gain at

---

[9]In this case, a multirate system may be worth considering as an alternative to reduce the system bandwidth to something closer to the signal bandwidth.

the output. For example, such peaks are not driven to their full potential gain by a typical sweep input, unless the sweep is very slowly changing. This suggests that a very narrow peak is not likely to cause an overflow in typical data and, therefore, not critical to accommodate in a scaling scheme. We conclude from this first-order analysis that if the frequency sampling is reasonably "finely spaced" and any frequency sweep testing is done with a reasonably "slowly varying" sweep, then overflow events due to sinusoidal input signals near a frequency response peak will be "very rare."

6.2.6 L_2 NORM SCALING

In a system where the signals are especially well-behaved, or if keeping a high SNR is much more important than avoiding overflow, it may be appropriate to consider an even less conservative scaling method. In this case, L_2 norm scaling may be useful. Since the L_2 norm is related to the signal energy, L_2 norm scaling can be considered to be restricting the signal energy, which is only very loosely related to the peak value of the signal.

To see how we can justify L_2 scaling, consider that the energy in the signal at node r can be written in terms of the L_2 norm in the frequency domain (Parseval's),

$$||x_r[n]||_2 = ||X_r(e^{j\omega})||_2 = ||X_0(e^{j\omega})H_{0,r}(e^{j\omega})||_2. \tag{6.27}$$

Since all p-norms satisfy the Cauchy-Schwarz Inequality[10] we know that

$$||x_r[n]||_2 \le ||X_0(e^{j\omega})||_2||H_{0,r}(e^{j\omega})||_2, \tag{6.28}$$

and therefore we can constrain the energy in every $x_r[n]$ by setting

$$s = \frac{1}{\max_r ||H_{0,r}(e^{j\omega})||_2} . \tag{6.29}$$

L_2 norm scaling is usually easier to compute in the time–domain, so we have

$$s = \frac{1}{\max_r ||h_{0,r}[n]||_2} = \frac{1}{\max_r \sqrt{\sum_n |h_{0,r}[n]|^2}} , \tag{6.30}$$

6.2.7   SCALE FACTOR COMPARISONS

We have introduced three types of scaling in order of increasing tolerance of overflow, but it can be said more precisely that for a given signal or impulse response, the value of the norms we have discussed can be shown to take on the corresponding order in their values. Ordering the commonly used norms in terms of the least restrictive (allows more overflow) to most restrictive (allows no overflow), gives the result shown in Table 6.1.

---

[10]The general form of Cauchy-Schwarz for discrete vectors is $||x[n]y[n]||_p \le ||x[n]||_p||y[n]||_p$. This is easy to prove using the identity $(\sum_n x[n])(\sum_n y[n]) = \sum_n \sum_m x[n]y[m]$, which can be visualized as a matrix of all the cross-terms of the product. A similar technique with a double integral works for the continuous function version. The reader should consider proving Cauchy-Schwarz in discrete-time as an exercise.

Table 6.1: Ordering commonly used norms.		
Least Restrictive $\leq$	$\cdots$ $\leq$	**Most Restrictive**
(a)	**(b)**	**(c)**
L_2 scaling	$L_{\infty\omega}$ scaling	L_{1t} scaling
$\|H_{0,r}(e^{j\omega})\|_2 = \|h_{0,r}[n]\|_2$	$\|H_{0,r}(e^{j\omega})\|_\infty$	$\|h_{0,r}[n]\|_1$

(a) L_2 scaling — gives the best SNR of the three if no overflow occurs. Even less restrictive scaling methods are available (see (6.18)), but not popular.

(b) $L_{\infty\omega}$ scaling — Full scale sinusoids can never overflow, but other signals could.

(c) L_{1t} scaling — Absolutely no overflow is possible, but this method produces the worst SNR of the three.

We don't need to calculate the impulse response at every filter node if we make careful choices of coefficient and signal representation format. For example, a $Q1.15$ coefficient multiplied by a $Q1.15$ signal and shifted to a $Q1.x$ result cannot overflow except in the pathological case of $-1 \times -1 = 1$. If we are willing to ignore this rare case, (saturating arithmetic makes it very unobtrusive) nearly all multiplier outputs can be ignored. If we do not wish to ignore even the one possible multiply overflow, we may be able to leave the results in double precision $Q2.30$ form where no overflow is possible and no precision is lost. Since the natural form of the double length product of any two Q formats cannot overflow this approach is practical even when coefficients are in $Q2.14$, etc. Overflow analysis can then be put off until the output of an addition.

Likewise, many potential addition overflow nodes can be ignored if wrapping arithmetic is used, since overflows at these nodes can self-correct, if the final sum does not overflow. With the use of wrapping arithmetic, only the final sum need be analyzed for overflow and appropriate scaling.

If saturation arithmetic is required, the range of the intermediate adder results can be increased enough to avoid possible overflows by shifting to a larger number of integer bits in the result. The loss of precision will increase the noise floor, but if double precision accumulators are used, the precision lost by a few bit shifts will be overwhelmed by the later effect of reducing the result back to single precision, and overflow analysis can again be deferred to a final node at the end of a chain of adders.

These simplifications mean that computation of impulse responses for scaling analysis rarely needs to be done anywhere other than at the beginning or end of a filter subsection where a set of adders combine products. This principle applies to a variety of filter structures such as Direct Form I, Direct Form II, and Direct Form II transposed, where successive additions occur without intermediate quantization.

6.2.8   CASCADED SECOND ORDER SECTIONS

When we implement cascaded second order sections, it is practical and beneficial to distribute the scaling throughout the filter instead of scaling for the entire filter with a single constant at the input. Distributed scaling is not available in a single Direct Form II section because there are no intermediate points in the flow-graph where a gain could be inserted without changing the frequency response. In a cascaded SOS filter distributing the scaling allows the signal to be larger in parts of the filter where the gain is lower, and therefore has the potential to raise the SNR. However, there are many parameters effecting SNR and scaling is only one of them.

If we consider the second order sections separately, a scale factor can be inserted in between each section.

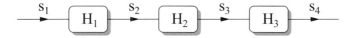

Figure 6.6: Inserting a multiplier between sections allows flexibility in scaling.

Then we can choose each scale factor according to the section following it.

- s_1 prevents overflow in H_1 internals and output.

- s_2 prevents overflow in H_2 internals and output.

- s_3 prevents overflow in H_3 internals and output.

- s_4 restores the result to its original level.

Note that the scaling computations of impulse responses at each node must be done sequentially. Since s_1 must be computed based on responses in H_1, the responses in H_2 cannot be determined until s_1 is fixed. Similarly, s_2 must be determined and fixed before calculations on nodes in H_3 can be performed to determine the value of s_3. Typically, a filter has been specified to have a pass-band gain of 1, or has some other overall gain requirement, and the scaling process will modify the overall gain by a factor of exactly $s_1 \cdot s_2 \cdot s_3$. Although the value of s_3 has been set to maximize the value of the output of H_3 without violating the overflow constraint, we may wish to restore the overall gain of the system by undoing the effects of the previous gain adjustments. In this case, $s_4 = \frac{1}{s_1 \cdot s_2 \cdot s_3}$. *This adjustment can result in overflows at the output even though the system has been scaled to prevent internal overflow.*

MATLAB's Scaling Process   Matlab provides support for scaling cascaded SOS filters in `tf2sos()` and related functions, in the `scale()` function, and in `fdatool`. As an example of the scaling process, we will describe the computations carried out by these Matlab functions.

Consider a fourth order cascaded SOS filter, using DFII sections. Scaling operations can be inserted at the input, between the sections, and at the output. If we are either careful (as discussed at the end of Section 6.2.7) or brave, we can ignore potential overflow everywhere but at the final

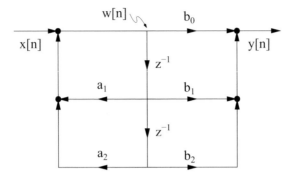

Figure 6.7: The Direct Form II section diagram shows the intermediate signal that is modified only by the poles.

adder for the feedback coefficients, and at the final adder for the feed-forward coefficients. Matlab's scaling support seems to assume that we have been careful.

Because the final adder for the feedback coefficients in a section is effectively the input to the delay line and feed-forward coefficients for that section, we can think of the response at that node as having passed through only the poles of the section, as in the intermediate signal $w_1[n]$ shown below. The signal $w_2[n]$ appears at the node of the output of the feed-forward adders, and can be thought of as the output of the first set of zeros (before any subsequent scaling). The $w_3[n]$ and $w_4[n]$ signals occupy similar positions in the second section.

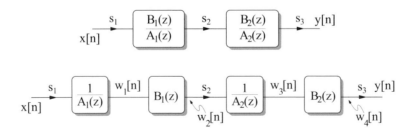

Figure 6.8: There is no opportunity to scale within a Direct Form II section, but the poles tend to have large gain requiring a scale correction.

In a Direct Form II section, usually $w_1[n]$ and $w_3[n]$ are the danger points, because poles tend to have high gain, and zeros have lower gain, although this is just a rule of thumb based on band select filter designs, and other design criteria could produce very different results. The common behavior of high pole gain and attenuation at the zeros is responsible for the main difficulty in IIR filter implementation. Since s_1 must reduce the input enough to prevent overflow at $w_1[n]$, it must also reduce the signal level at $w_2[n]$ which is already reduced by the attenuation of the zeros in $B_1(z)$. This double attenuation is difficult to avoid because scaling cannot be adjusted in between $\frac{1}{A_1(z)}$ and $B_1(z)$, and the effect is to cause a low SNR in the $w_2[n]$ signal due to the low signal level. When poles are very close to the unit circle and high gains result, the need for wide dynamic range sometimes drives designers to use extra (possibly double) precision for a section.

Now that the nodes for overflow analysis have been identified, s_1 can be determined based on the L_{1t}, $L_{\infty\omega}$, or L_2 norms of the impulse responses at the nodes $w_1[n]$ and $w_2[n]$. When s_1 is being calculated, its value is treated as 1 so that any potential to overflow will be apparent. As noted, $w_1[n]$ will usually be the node causing the maximum value. With s_1 set, the impulse responses at $w_3[n]$ and $w_4[n]$ can be determined using $s_2 = 1$. The resulting norms used to calculate the s_2 necessary to set the maximum norm back to 1. Matlab always sets the value of $s_3 = \frac{1}{s_1 \cdot s_2}$ to restore the original filter gain.

One might reasonably imagine that since Matlab's dfilt.df2sos objects include scale value parameters defined in the positions of s_1, s_2, and s_3, that the scaling process would set these values according to the calculations above. However, upon observation one usually finds s_2 to be 1, and often $s_3 = 1$ as well. This is because s_2 can usually be factored into the previous b coefficients, saving both a multiply and the associated quantization step. All that is required is replacing $b_{0,1} \rightarrow s_2 b_{0,1}$, $b_{1,1} \rightarrow s_2 b_{1,1}$, and $b_{2,1} \rightarrow s_2 b_{2,1}$. There are no previous coefficients to factor s_1 into, so it usually remains as calculated. The value of s_3 depends on several factors. Usually, s_1 is small to prevent overflow and s_3 is large undo the effects of s_1. Although s_3 could always be factored into the $b_{i,2}$ coefficients, the resulting coefficient values could become quite large, and Matlab can limit the coefficient value to a maximum value, typically 2, so that $Q2.14$ coefficients could be used. If the limit on the amplitude of the b coefficients overrides the policy of factoring in s_3, the remainder will appear as a separate multiplier.

It is well known that using scale values with values that are powers of two allows hardware implementation as a shift instead of a multiply. However, powers of two can also be useful because of their exact representation. For example, in Matlab's dfilt.df2sos objects, all the scale values must be represented using the same Q format even though the first scale value is usually small and the last scale value is usually large (often nearly the reciprocal of the first scale value). This constraint is obviously to simplify implementation, but it also can cause precision problems.

As an example, consider a system where the first scale value, $s_1 = 1/100$, and the last scale value, $s_3 = 65$. If a 16 bit word is used, we will have to choose $Q8.8$ for our scale values to avoid overflow in s_3, but this only leaves 8 bits of precision to represent the binary repeating fraction needed for s_1. If we quantize s_1 to $Q8.8$, the result rounds to 0.0117, not 0.01, so the filter's gain is off by 17%, and overflows may occur that were intended to be suppressed. The problem is that the large value gets all the precision, and the smaller one gets almost none. If we can afford to scale down more, we can shift the gain of s_1 to a power of two, which needs no extra bits, and let the precision of the larger value take care of setting the filter gain precisely. In this case we might use $s_1 = 1/128$ and $s_3 = 65 \cdot \frac{128}{100} = 83.1992$. A similar solution that requires less scaling down would be to simply round s_1 down to the nearest representable value, and then correct s_3 for the difference in gain. In either case, s_3 can be represented with far more precision than s_1.

Ordering Poles and Zeros   In addition to adjusting the gain through the filter, we can change the noise performance by changing the order of the poles and zeros in each section. No simple analytical solution is available to resolve the question of which ordering is best, other than simply testing all the possibilities, or with some cleverness, dynamic programming [25] to only test the best possibilities. For high order filters, this can be prohibitive, since for L sections there are $L!$ possible section orders, and $L!$ possible pole-zero groupings.

A simple and effective rule of thumb is available for grouping the poles and zeros together in each section. Since we wish to minimize peaks in gain for each section, it makes sense to choose the highest radius pole and group it with the nearest zero first, proceeding to lower and lower radius poles. This minimizes (or nearly) the peak gain in each section, which tends to reduce the need for scaling.

Unfortunately, the best section ordering choices are less obvious. Jackson [13] suggests that noise power is reduced in an $L_{\infty\omega}$ normed filter when the poles are ordered from farthest from the unit circle to nearest ("up" order in Matlab scaling functions), and that peak noise spectral density is reduced when the poles are ordered from closest to the unit circle to farthest ("down" order in Matlab) in an L_2 scaled filter. One of these two orderings is usually a good choice. However, Schlicthärle [25] describes a more complicated rule of thumb, which may improve on the simple "up" and "down" choices.

Another complication is that various orderings may change probability of overflow as well as the noise power and noise spectrum at the output. We may not wish to obtain a high SNR by accepting a high probability of overflow, so it is often necessary to test a system with actual signal data and determine the SNR at a signal level that avoids overflow problems. The final SNR for a given signal at the peak amplitude without overflow can depend on the scaling method, pole and zero order, and the signal itself. From experience, we have found that reasonably good performance can be obtained by choosing a scaling method consistent with our desired probability of overflow (and willingness to give up SNR), and then checking the SNR for the up and down pole orderings for Direct Form I and Direct Form II filter structures. In an example of running speech data through a 4th order low pass elliptic filter with 16 bit data, it is reasonable to expect as high as 70 dB SNR, or much lower depending on the choice of order and scaling.

A key piece of intuition for analyzing these structures is that if we preserve the output gain, any scaling down of the intermediate signal to avoid overflow will have to be accounted for by increasing gain in the "downstream" portion of the filter, and the compensating gain will necessarily amplify any noise injected at the scaled node, and possibly later nodes as well. For this reason, it is especially helpful to avoid large gains at nodes where quantization occurs.

Dattorro [3] recommends the use of Direct Form I structures so that accumulation (at double precision) for both poles and zeros can be done at a single node, and wrapping arithmetic can be used to avoid any intermediate overflows. Saturation and quantization can be applied to the result before passing it on to the next section at the multiplier input precision. Note that this approach always quantizes results where the poles are partially canceled by the zeros, and thus overall gain tends to be small, reducing the need for scaling.

Dattorro also suggests the use of truncation error feedback (also discussed in Mitra [20]). This clever technique uses either approximate or exact copies of the feedback coefficients on the quantization error to remove the gain effects of the poles on the quantization error, thus reducing its power dramatically. The computational cost is significant, but if carefully implemented, all the multiplies can be carried out with similar word-lengths to the signal data (though with smaller fractional parts).

6.3   DESIGN EXAMPLE

The principles behind the calculation of scale factors and noise power at the output are fairly simple, but can seem rather abstract when described in general as above. This example is intended to make the ideas more concrete by applying the principles to a specific system. Because we hope to show the origin of the calculations, some detail will be necessary, and if these calculations were carried out by hand for every node they would be rather tedious. Fortunately, computer tools allow us to rapidly compute these quantities where needed, but it is still very illuminating to examine the structure of an example. [11]

6.3.1   SCALING CALCULATIONS

Consider for this example the filter designed by the Matlab command `[b,a] = ellip(4,1,20,0.2)`. The resulting 4th order bilinear elliptic filter should have a pass-band ripple of 1 dB, a stopband attenuation of at least 20 dB, and a cutoff frequency of $\omega_c = 0.2\pi$. The resulting coefficients are approximately

```
b =

    0.1082    -0.2499    0.3455    -0.2499    0.1082


a =

    1.0000    -2.9870    3.7726    -2.2866    0.5707
```

and can be represented by the transfer function

$$H(z) = \frac{0.1082 - 0.2499z^{-1} + 0.3455z^{-2} - 0.2499z^{-3} + 0.1082z^{-4}}{1 - 2.9870z^{-1} + 3.7726z^{-2} - 2.2866z^{-3} + 0.5707z^{-4}}. \tag{6.31}$$

[11]See the website for source code in `ch6example.m`.

This can easily be factored (`tf2sos()`) into a form convenient for cascaded second order sections implementation and becomes

$$H(z) = 0.1082 \frac{(1 - 0.7798z^{-1} + z^{-2})}{(1 - 1.4199z^{-1} + 0.6064z^{-2})} \frac{(1 - 1.5290z^{-1} + z^{-2})}{(1 - 1.5671z^{-1} + 0.9412z^{-2})}. \tag{6.32}$$

From this cascaded form, we can see that there are many choices for pole and zero ordering, and an infinite number of ways to distribute the initial constant over the two sections by multiplying a constant times the numerator coefficients. By noting that the a_{2k} coefficient in each section is also the pole radius squared (see derivation in (5.19) and (5.21)), we can see that the first section has a smaller pole radius than the second. Since the pole radius increases from left to right, Matlab refers to this ordering as 'UP' order. The zeros are paired with the poles by choosing the zeros closest to the highest radius pole pair (to maximize gain cancellation), and then repeating for each lower radius pole pair.

If we wish to restrict the flow-graph to two five-coefficient sections (not counting the $a_{0k} = 1$ terms) there is no place to implement the gain separately and it must be combined with the numerator coefficients. Without applying a norm-based scaling algorithm, the simplest thing to do is to combine all of the gain constant with the first section's numerator, producing the system function

$$H(z) = \frac{(0.1082 - 0.0844z^{-1} + 0.1082z^{-2})}{(1 - 1.4199z^{-1} + 0.6064z^{-2})} \frac{(1 - 1.5290z^{-1} + z^{-2})}{(1 - 1.5671z^{-1} + 0.9412z^{-2})}. \tag{6.33}$$

Now suppose we wish to scale the system using the $L_{\infty\omega}$ norm (max frequency response). If we assume a Direct Form II section, the poles will be implemented first, and the scaling values can only appear before, after, or between sections. Scaling multiplies occurring after a section may be combined with the preceding zeros.

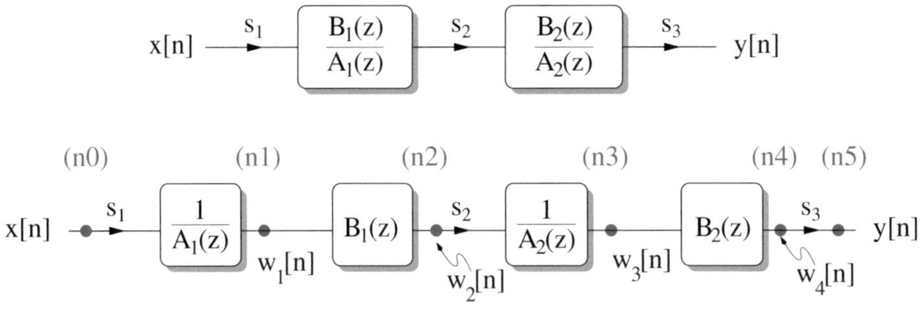

Figure 6.9: The flow-graph with intermediate signals shown between the poles and the zeros. Node numbers referred to in the text are labeled as (n0) through (n5).

In the current example, we must choose s_1 to prevent overflow at the intermediate signal nodes $w_1[n]$ and $w_2[n]$. Note this implicitly assumes that the arithmetic will be arranged so that overflows at the intervening multiplies and adds will be impossible. Intermediate overflows can be avoided by using the appropriate Q-Format with a double precision accumulator, or by allowing wrapping.

If we now use a default value of $s_1 = 1$ and define

$$H_{0,1}(z) = \frac{1}{A_1(z)} \quad \text{and} \quad H_{0,2}(z) = \frac{B_1(z)}{A_1(z)} \tag{6.34}$$

then s_1 for the $L_{\infty\omega}$ norm can be computed as

$$s_1 = \frac{1}{\max\limits_{\omega,k=1,2} |H_{0,k}(e^{j\omega})|} \tag{6.35}$$

Although this calculation can be numerically sensitive when there are narrow peaks in the magnitude response, the results are useful even if only approximate. For the example at hand, we obtain

$$\max_\omega |H_{0,1}(e^{j\omega})| = 6.1826$$

and

$$\max_\omega |H_{0,2}(e^{j\omega})| = 0.7496$$

so that

$$s_1 = 1/6.1826 = 0.1617.$$

This scaling adjustment should prevent a pure sinusoid from causing overflow in the first section, but it is approximate due to frequency sampling in the DFT. Now s_2 can be adjusted to prevent sinusoid overflow in the second section, but the value of s_1 must be included. It should come as no surprise that the node following the poles has the maximum gain since frequency selective filters often place poles near the unit circle and then offset the resulting gain with attenuation in the zeros. Therefore the node following the poles is often the limiting case for overflow.

Starting with a default value of $s_2 = 1$, we will define

$$H_{0,3}(z) = s_1 \frac{B_1(z)}{A_1(z)A_2(z)} \quad \text{and} \quad H_{0,4}(z) = s_1 \frac{B_1(z)B_2(z)}{A_1(z)A_2(z)} \tag{6.36}$$

so that s_2 can be computed as

$$s_2 = \frac{1}{\max\limits_{\omega,k=3,4} |H_{0,k}(e^{j\omega})|} \tag{6.37}$$

For the example at hand, we obtain

$$\max_\omega |H_{0,3}(e^{j\omega})| = 1.6241$$

and

$$\max_\omega |H_{0,4}(e^{j\omega})| = 0.1617$$

so that

$$s_2 = 1/1.6241 = 0.6157.$$

This scaling adjustment will prevent any pure sinusoid from causing overflow in the second section.

Traditionally, the purpose of s_3 is not to prevent overflow (s_2 would be sufficient for this function) but to restore the output gain to the designed value. Keep in mind that the resulting output is **not** protected from overflow under these conditions, and the overall system design must be appropriately scaled. Restoring the original gain simply requires that $s_3 = \frac{1}{s_1 s_2}$. In our example,

$$s_3 = 10.0413.$$

To explicitly show the scaling values and scaled coefficients, we could represent the scaled system in the form

$$H(z) = (0.1617)\frac{(0.1082 - 0.0844z^{-1} + 0.1082z^{-2})}{(1 - 1.4199z^{-1} + 0.6064z^{-2})}(0.6157)$$
$$\frac{(1 - 1.5290z^{-1} + z^{-2})}{(1 - 1.5671z^{-1} + 0.9412z^{-2})}(10.0413). \quad (6.38)$$

Performing these operations in Matlab will produce similar but not identical results, because Matlab's functions make a few adjustments for improved implementation. The first adjustment is to note that the s_2 multiply immediately follows the $B_1(z)$ coefficients, and so both the required multiply and any quantization can be eliminated by combining the two operations into $s_2 B_1(z)$. It is also possible to combine s_3 and $B_2(z)$, but that result would produce unusually large coefficient values. By default, Matlab limits the maximum coefficient magnitude in $B_2(z)$ to 2 (a reasonable choice assuming $Q2.14$ representation of the coefficients), and leaves the rest of the gain in the s_3 multiplier. These adjustments lead to the explicit system function form

$$H(z) = (0.1617)\frac{(0.0666 - 0.0520z^{-1} + 0.0666z^{-2})}{(1 - 1.4199z^{-1} + 0.6064z^{-2})}(1)$$
$$\frac{(1.3080 - 2z^{-1} + 1.3080z^{-2})}{(1 - 1.5671z^{-1} + 0.9412z^{-2})}(7.6768) \quad (6.39)$$

which is the same result as obtained from the Matlab [12] code below.

```
[b,a]=ellip(4,1,20,.2);           % create an example filter
sosm = tf2sos(b,a);              % convert to second order sections
Hd = dfilt.df2sos(sosm);         % create a df2 dfilt object
scale(Hd,'Linf','sosreorder','up'); % scale the filter
Hd.scalevalues                   % report the scalevalues
Hd.sosmatrix                     % report the filter coefficients
```

[12]Using R2008b, with the Filter Design Toolbox. Other versions could make slightly different implementation assumptions and produce slightly different gain distributions.

```
scv =

    0.1617
    1.0000
    7.6768


sosmat =

    0.0666   -0.0520    0.0666    1.0000   -1.4199    0.6064
    1.3080   -2.0000    1.3080    1.0000   -1.5671    0.9412
```

Example 6.3.1 (L_{1t} Norm). If you understand the preceding example thoroughly, you should be able to produce a similar justification for the results of the filter design scaled with the L_{1t} norm. To check your results, simply replace 'Linf' with 'L1' in the Matlab code above.

6.3.2   OUTPUT NOISE CALCULATIONS

In the previous subsection, we concerned ourselves with the effect of the filter from the input to the node at which we evaluated the norm. Now, however, we will be concerned with the sources of quantization noise, and the effect of the filter from the node at which noise is added to the output.

Although the implementation details of a given filter can vary dramatically with certain design decisions and with the constraints of the computational architecture, we must make a specific set of implementation assumptions for our example. Here we will use a set of restrictions fairly common to 16 bit microprocessors. We will assume multipliers with 16 bit inputs and 32 bit outputs, and adders with 32 bit accumulators. These assumptions imply that every product must be re-quantized before it can be multiplied by a new coefficient, but that sums can be implemented on 16 or 32 bit quantities. We will also assume that the system expects a 16 bit input and output. If the default signal is represented in $Q1.15$ format, the noise power injected at each quantizer (shown in Fig. 6.10) will be $\sigma_e^2 = \frac{2^{-30}}{12} = 7.76 \times 10^{-11}$. We will assume the quantization noise is zero mean, white, uncorrelated with other sources, and uniform distributed as usual. Of course, the filter after the injection node will shape the noise spectrum and change the power, so we must compute the noise spectrum and power carefully.

We will define the transfer function from the left-most noise source to the filter output as

$$H_{1,5}(z) = \frac{B_1(z)}{A_1(z)} s_2 \frac{B_2(z)}{A_2(z)} s_3 \tag{6.40}$$

since the entire filter is involved except for s_1. If you are uncertain, be sure to take a moment convince yourself that noise injected at (n1) is modified by the poles of $\frac{1}{A_1(z)}$. The transfer function shaping

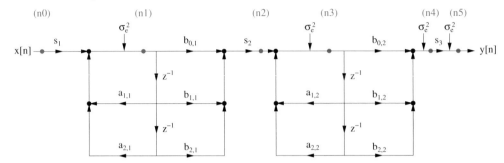

Figure 6.10: Using the quantization noise model, noise sources are injected at each point in the filter where precision is lost. The node numbers mentioned in the text are labeled as (n0) to (n5).

the second noise source is defined as

$$H_{3,5}(z) = \frac{B_2(z)}{A_2(z)} s_3 \tag{6.41}$$

and the last two sources are shaped by the transfer functions

$$H_{4,5}(z) = s_3 \quad \text{and} \quad H_{5,5}(z) = 1, \tag{6.42}$$

respectively.

Using the relationships developed earlier in Section 6.1 we can now write the output noise power as

$$\sigma_{out}^2 = \sigma_e^2 \left(\sum_{n=0}^{\infty} h_{1,5}^2[n] + h_{3,5}^2[n] + h_{4,5}^2[n] + h_{5,5}^2[n] \right), \tag{6.43}$$

where $h_{k,5}[n]$ is simply the impulse response at the system output of the partial filter modifying noise injected at the kth node. This quantity would be difficult to compute analytically for all but the simplest filters, but we can easily approximate the result from thousands of impulse response values in Matlab. Although we have assumed equal noise power for each quantization operation, the power at the output for each differs substantially. For this example filter we find that the output quantization noise power, σ_k^2, due to the source at node k is

$$
\begin{aligned}
\sigma_1^2 &= 5.5529 \times 10^{-10} \\
\sigma_3^2 &= 9.5023 \times 10^{-9} \\
\sigma_4^2 &= 4.5738 \times 10^{-9} \\
\sigma_5^2 &= 7.7610 \times 10^{-11}
\end{aligned}
$$

with the combined noise power at the output producing

$$\sigma_{tot}^2 = 1.4709 \times 10^{-8}.$$

Note that despite the bandwidth reduction by the filter, all the internal sources contribute more power to the output than they input because they are amplified by the scaling correction gain at the end. Only the final source remains unamplified. The combined gain of the poles in the second section and the gain correction make the source at node 3 the dominant one. These adjustments should remind the reader of the rule that when output gain is fixed, any attenuations to prevent overflow must be corrected in later sections and such gain corrections will amplify subsequent noise sources.

Dattorro [3] points out that a Direct Form I structure often has an advantage in that when the outputs of a pole subsection and zero subsection are combined, wrapping arithmetic can be used so that no overflow is possible in intermediate results. Therefore, the gain of the poles does not need to be protected by attenuation until after it is partially canceled by the nearby zeros. Reduced need for scaling means reduced noise amplification.

For the wide range of input signals that meet the criteria for treating quantization as noise, noise power and spectra at the output will be quite predictable. However, computations of signal-to-noise ratio (SNR) or probability of overflow are highly dependent on the specific input signal, particularly its crest factor and distribution of values. We will consider SNR for a particular signal later in this example, but first we should compute the noise spectrum at the output. After all, the spectral shape of the noise can have major impact on both the subjective and quantitative performance of the system.

6.3.3   NOISE SPECTRUM CALCULATIONS

Since each quantizer supplies white noise with a power of σ_e^2, the input power spectrum is $P_e(\omega) = \sigma_e^2$ and the output power spectrum for each source is just

$$P_k(\omega) = \sigma_e^2 |H_{k,5}(e^{j\omega})|^2. \tag{6.44}$$

Since uncorrelated power spectra add, the total output spectrum is

$$P_{out}(\omega) = \sigma_e^2(|H_{1,5}(e^{j\omega})|^2 + |H_{3,5}(e^{j\omega})|^2 + |H_{4,5}(e^{j\omega})|^2 + |H_{5,5}(e^{j\omega})|^2). \tag{6.45}$$

Figures 6.11 and 6.12 use a two sided definition of power spectrum, $P_{total} = \frac{1}{2\pi} \int_{-\pi}^{\pi} P(e^{j\omega}) d\omega$, but only one side is shown.

Note that the reader verifying these results by simulation should be careful to separate coefficient quantization error from product quantization error since the mechanisms of error are different, and only product quantization noise is shown in the figures. This separation is easily accomplished by using the quantized coefficient values in both the "ideal arithmetic" simulation and in the quantized simulation.

6.3.4   SNR CALCULATIONS

Signal-to-noise ratio calculations inherently depend on the input signal power and the power spectrum of the input. It would be easy to choose a white input signal for comparison purposes but a

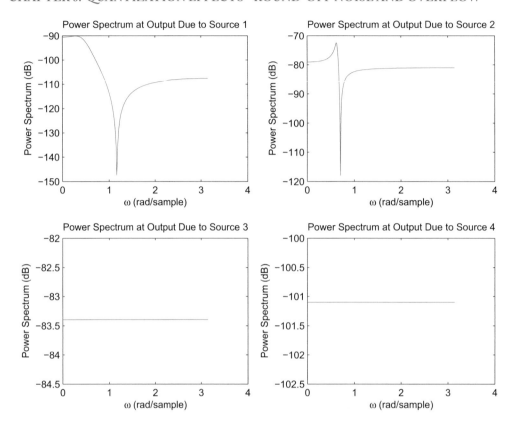

Figure 6.11: The internal white noise sources are individually shaped and amplified by the filter they pass through to the output. The fourth source is neither shaped nor amplified.

constant spectrum is slightly oversimplified, and a white signal does not lend itself to audio verification for intuition. So, instead, we will use a recorded speech signal as the input signal to emphasize the nonuniform spectrum and allow for audio verification of the results. The input signal file, `began.wav`, has been provided on the website.

We have already found that the average noise power due to quantization sources is $\sigma_{out}^2 = 1.4709 \times 10^{-8}$. The average power of the filtered signal is $P_{speech} = 0.0576$, so the resulting SNR will be

$$\text{SNR} = 3.9148 \times 10^6 = 65.9271 \text{ dB}.$$

Of course, both the signal power and the noise power can be expressed as a function of frequency using spectral estimation for the signal and the noise spectrum computations shown above. Since noise in the pass-band is more difficult to remove later, we may want to determine the SNR over a restricted band, and spectral information makes this possible. Figure 6.13 shows both the signal

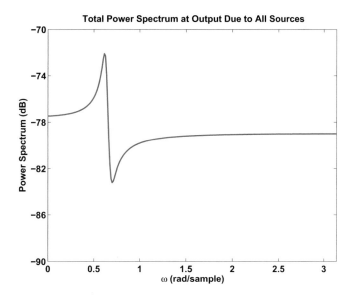

Figure 6.12: The combined output noise spectrum shows the contributions of all the sources.

and the noise spectra for comparison. If we sum[13] the signal and noise power up in, for example, just the pass-band $(0 \leq \omega \leq 0.2\pi)$ we can find that the pass-band SNR is

$$SNR_{passband} = 71.8569 \text{ dB}.$$

This means that further high precision filtering could produce a higher SNR later in the signal path.

6.3.5   COMPARING DIFFERENT CHOICES

When designing a filter for a system, the bottom line comes down to which design gives the best performance. Often the designer will have a representative set of test data for which fewer than some maximum number of overflows must occur. If the test data set is large enough to be truly representative, then the overflow and SNR results outweigh the choices dictated by a particular scaling method.

In the example comparing various scaling choices below we will find a case where the unscaled filter produces a high SNR, but extra attenuation must be included to prevent overflow. In other cases we will find that the standard scaling methods include too much attenuation and the output signal level can be increased by reducing the input attenuation. Keep in mind that these input gain adjustments do not preserve the output signal level, and consequently change signal power and SNR.

If we want to compare the best possible SNR from a given set of design choices for section order and scaling method, it makes sense to check to see if the test data can be amplified without

---

[13]Be careful not to do this operation in dB!

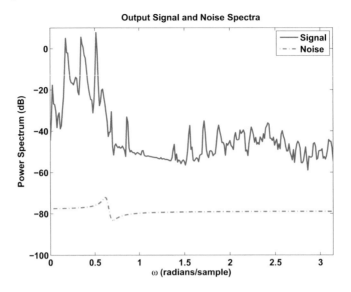

Figure 6.13: Comparing the signal and noise spectra shows how the system behaves differently in the pass-band and the stopband.

causing overflow or must be attenuated to prevent overflow and incorporate the best input gain for each design in the SNR measurements. Since a quantized signal with values near full-scale cannot be amplified gain adjustments are best accomplished by adjusting the filter's input scaling coefficient s_1.

Figure 6.14, shows a comparison of twelve different design choices. The design choices evaluated include the two Direct Form filter structures (DF1 and DF2),[14] section order (up and down), and scaling method ($L_{\infty\omega}$, L_{1t}, and unscaled). The unscaled designs apply all the attenuation to the zeros of the section with the poles closest to the unit circle, and the zeros of the other section retain a $b_{0k} = 1$.

Since we are measuring overflow for a particular signal and not a general class of signals, it should not be surprising that a scaling choice (the unscaled, but adjusted DF1-down choice) can be found that achieves better SNR than either the $L_{\infty\omega}$ or the L_{1t} methods. Note that the max gain of about 0.75 means that extra attenuation was needed to prevent overflow, but even with reduced signal gain, the SNR was still higher that the other methods. Similarly, we see that all four of the systematically scaled Direct Form II design choices were overly conservative and the input could be scaled up by a factor of almost 1.2 without causing overflow. However, even with their improved signal power, they remain inferior in SNR to all of the Direct Form I design choices.

[14]Direct Form I (DF1) and Direct Form II (DF2)

When we look at Fig. 6.14 for patterns among the filter structure, scaling, and section order choices we see that the Direct Form I structures are generally preferable to the Direct Form II structures as predicted. Recall that Direct Form I requires double the storage of a Direct Form II structure, so this benefit is not free. Note also that in every case the $L_{\infty\omega}$ scaled filter has a higher SNR than the complementary L_{1t} scaled filter, and finally, that up ordering produces higher SNR than down ordering in most cases. When large representative data sets and exhaustive testing are not feasible, these rough rules of thumb can be good guides for starting points.

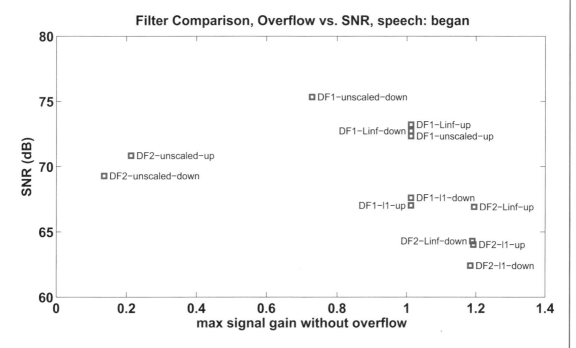

Figure 6.14: A comparison of several combinations of scaling and section ordering. The unscaled designs lump all the attenuation into the section with poles farthest from the unit circle. The other section uses a b_0 coefficient of 1. Other signals and other filters will produce different results.

CHAPTER 7

Limit Cycles

Limit cycles, unlike the quantization effects discussed previously, cannot be analyzed using simple linear approximations. There are two basic types of limit cycles

- overflow limit cycles, and

- rounding oscillation limit cycles.

Overflow limit cycles occur primarily when wrapping arithmetic (two's complement, etc.) produces an overflow result that is far from the correct value. The error injects a large transient into the system, and if feedback allows the transient response to cause another overflow, the oscillation can become self-sustaining. Fortunately, the use of saturating arithmetic can preclude overflow limit cycles in second order direct form filters [5]. Dattorro [3] suggests the use of saturation only for final addition results in each section so that the benefits of wrapping intermediate results (and perhaps reduced hardware cost) are available.

Rounding oscillation limit cycles occur when the error introduced by rounding (or truncation, etc.) is enough to prevent the system from decaying to zero when no input is present. Note that even a non-zero constant output with zero input is a limit cycle effect, if not an oscillation. A similar oscillation effect can prevent decay to a constant output with a constant input. Generally these effects have very small amplitudes (often only the least significant bit changes), and increased precision can always reduce limit cycle amplitude. Because increased precision can be expensive or impractical, numerous other methods exist for mitigating rounding limit cycles. The most common of these methods is magnitude truncation arithmetic.

To see these effects, consider a few examples. We can set up a simple system implemented in $Q1.3$ arithmetic for easy hand checking. Assume rounding back to $Q1.3$ occurs after every mathematical operation. In the system whose difference equation is $y[n] = -\frac{5}{8}y[n-1] + x[n]$ with zero input $x[n] = 0$ and an initial condition of $y[0] = \frac{3}{8}$, the output will be

$$y[n] = \frac{3}{8}, \quad -\frac{1}{4}, \quad \frac{1}{8}, \quad -\frac{1}{8}, \quad \frac{1}{8}, \quad -\frac{1}{8}, \quad \frac{1}{8}, \quad -\frac{1}{8}, \ldots$$

Limit cycles occur in this system in the least significant bit of any word size. We can even change the sign of the pole coefficient or initial condition and produce variations on the limit cycle effect. Figure 7.1 shows the four possible sign variations and their resulting outputs. The thicker lined plot shows the quantized results, while the thinner lined plot shows the unquantized result.

Intuitively, rounding is causing the energy in the signal to increase when it should be decreasing, since rounding can increase the amplitude of the signal by rounding "up." One partial solution is

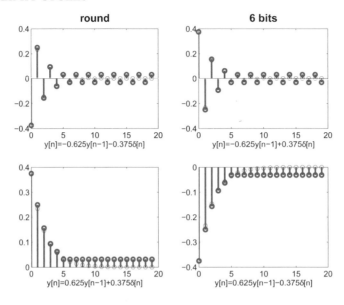

Figure 7.1: The quantized (6 bit arithmetic with rounding) outputs of four similar systems exhibiting limit cycle behavior are shown with thick circles. The thin circles indicate the unquantized system behavior. The difference equation of each system is shown below the plot.

to avoid rounding and use truncation (often with the added benefit of simpler hardware) instead. The results of the truncation, or "floor" approach are shown in Figure 7.2. Figure 7.2 shows that although truncation removes the limit cycle effect in many cases, the behavior of rounding toward negative infinity can increase the amplitude of a negative signal, and therefore may add energy to one of the example systems. Only magnitude truncation, implemented by the "fix" function in Matlab can reduce amplitude for every signal, since it rounds toward zero. Of course, the results are less precise when using magnitude truncation since the least significant bit is less accurate. As always, more precision can be achieved by increasing the wordlength. Figure 7.3 shows the results of using magnitude truncation, and by comparison with the previous figures, the reduced accuracy is apparent.

The suppression of limit cycles is a broad topic with all the complexity to be expected in a nonlinear system behavior. The most basic tools of saturation arithmetic and magnitude truncation rounding have been discussed here because they are sufficient for many systems, though certainly not all. However, there is a wide literature on the analysis and suppression limit cycles that is not appropriate for the scope of this text.

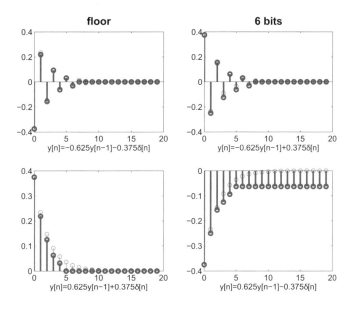

Figure 7.2: The quantized (6 bit arithmetic with "floor" rounding, or truncation) outputs of four similar systems are shown. The limit cycle behavior of three of the systems is suppressed.

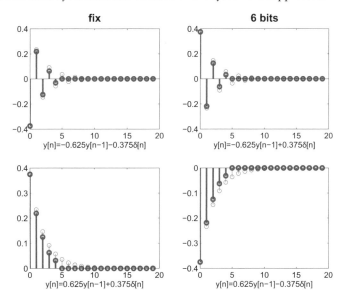

Figure 7.3: The quantized (6 bit arithmetic with "fix" rounding, or magnitude truncation) outputs of four similar systems are shown. The limit cycle behavior of all four systems has been suppressed, at the expense of reduced accuracy.

Glossary

A/D	Analog to Digital
BER	Bit Error Rate
CD	Compact Disc
CDF	Cumulative Density Function
CF	Crestfactor
CIC	Cascaded Integrator Comb
dB	decibel
DC	Direct Current (or constant)
DF1	Direct Form 1
DF2	Direct Form 2
DF2T	Direct Form 2 Transposed
DFT	Discrete Fourier Transform
DSP	Digital Signal Processing
FDATool	Filter Design and Analysis Tool
FFT	Fast Fourier Transform
FPGA	Field Programmable Gate Array
FIR	Finite Impulse Response
GUI	Graphical User Interface
IEEE	Institute of Electrical and Electronics Engineers
IIR	Infinite Impulse Response
LTI	Linear Time–Invariant
PDF	Probability Density Function
PRR	Peak to RMS Ratio
QAM	Quadrature Amplitude Modulation
LMS	Least Mean Squared
LSB	Least Significant Bit
μP	microprocessor
MSB	Most Significant Bit
RMS	Root Mean Square
SNR	Signal to Noise Ratio
SOS	Second Order Sections

Bibliography

[1] R. Allen and K. Kennedy. *Optimizing Compilers for Modern Architectures*. Elsevier / Morgan Kaufmann, 2002.

[2] Jeff Bezanson. Understanding floating point numbers. CProgramming.com. [Online] Available Aug. 2009:
`http://www.cprogramming.com/tutorial/floating_point/understanding_floating_point_representation.html`.

[3] Jon Dattorro. The implementation of recursive digital filters for high-fidelity audio. *Journal of the Audio Engineering Society*, 36(11):851–878, November 1988. See also `http://www.stanford.edu/~dattorro/HiFi.pdf` for some corrections.

[4] Günter F. Dehner. Noise optimized IIR digital filter design - tutorial and some new aspects. *Elsevier Signal Processing*, 83(8):1565–1582, 2003. DOI: 10.1016/S0165-1684(03)00075-6

[5] P. M. Ebert, J. E. Mazo, and M. G. Taylor. Overflow oscillations in digital filters. *Bell System Technical Journal*, 48(11):2999–3020, November 1969.

[6] David Goldberg. What every computer scientist should know about floating-point arithmetic. *ACM Computing Surveys*, 23(1):5–48, March 1991. [Online] Available Aug. 20009: DOI: 10.1145/103162.103163

[7] John L. Hennessy and David A. Patterson. *Computer Architecture: A Quantitative Approach*. Morgan-Kaufmann, third edition, 2002.

[8] IEEE Task P754. IEEE Std 754-2008, IEEE Standard for Floating-Point Arithmetic, 2008. [Online] Available Aug. 2009:
`http://ieeexplore.ieee.org/servlet/opac?punumber=4610933`.
DOI: 10.1109/IEEESTD.2008.4610935

[9] International Telecommunications Union. ITU-T Recommendation G.711 (STD.ITU-T RECMN G.711-ENGL 1989), 1989. [Online] Available Aug. 2009: `http://www.itu.int/rec/T-REC-G.711/e`.

[10] L. B. Jackson. On the interaction of roundoff noise and dynamic range in digital filters. *Bell System Technical Journal*, 49(2):159–184, February 1970.

[11] L. B. Jackson. Beginnings: The first hardware digital filters. *IEEE Signal Processing Magazine*, pages 55–81, November 2004. DOI: 10.1109/MSP.2004.1359141

[12] L. B. Jackson, J. F. Kaiser, and H. S. McDonald. An approach to the implementation of digital filters. *IEEE Transactions on Audio and Electroacoustics*, AU-16(3):413–421, September 1968. DOI: 10.1109/TAU.1968.1162002

[13] Leland B. Jackson. *Digital Filters and Signal Processing*. Kluwer, third edition, 1996.

[14] Brian W. Kernighan and Dennis M. Ritchie. *C Programming Language*. Prentice Hall, second edition, 1988.

[15] Donald E. Knuth. *The Art of Computer Programming*, volume 2 / Seminumerical Algorithms. Addison Wesley, third edition, 1998.

[16] K. A. Kotteri, A. E. Bell, and J. E. Carletta. Quantized FIR filter design using compensating zeros. *IEEE Signal Processing Magazine*, pages 60–67, November 2003. DOI: 10.1109/MSP.2003.1253556

[17] Richard G. Lyons. *Understanding Digital Signal Processing*. Prentice Hall, second edition, 2004.

[18] James H. McClellan, Ronald W. Schafer, and Mark A. Yoder. *DSP First: A Multimedia Approach*. Prentice Hall, 1998.

[19] James H. McClellan, Ronald W. Schafer, and Mark A. Yoder. *Signal Processing First*. Pearson Prentice Hall, 2003.

[20] Sanjit K. Mitra. *Digital Signal Processing*. McGraw-Hill, third edition, 2005.

[21] Gordon E. Moore. Cramming more components onto integrated circuits. *Electronics*, 38(8), April 1965. [Online] Available Aug. 2009: DOI: 10.1109/JPROC.1998.658762

[22] A. V. Oppenheim. Realization of digital filters using block floating-point arithmetic. *IEEE Transactions on Audio and Electroacoustics*, AU-18(2):130–136, June 1970.

[23] A. V. Oppenheim and R. W. Schafer. *Discrete-Time Signal Processing*. Prentice Hall, third edition, 2010.

[24] Athanasios Papoulis. *Probability, Random Variables, and Stochastic Processes*. McGraw-Hill, second edition, 1984.

[25] Dietrich Schlichthärle. *Digital Filters*. Springer, 2000.

[26] John R. Treichler, C. Richard Johnson, Jr., and Michael G. Larimore. *Theory and Design of Adaptive Filters*. Prentice Hall, 2001.

[27] Neil H. E. Weste and David Harris. *CMOS VLSI Design: A Circuits and Systems Perspective*. Addison Wesley, third edition, 2004.

[28] Bernard Widrow and István Kollár. *Quantization Noise*. Cambridge University Press, 2008.